Praise

The War and Environment Reader

"Human intelligence, in its unique glory, has contrived two mechanisms to destroy us and much else: nuclear weapons and environmental catastrophe, closely linked. It has yet to contrive the cure for the lethal malady, a task that must be undertaken without delay. The essays collected here explore the depths and imminence of the crises, and outline directions that must be pursued, and urgently."

—NOAM CHOMSKY
Institute Professor & Professor of
Linguistics emeritus, MIT

"We don't really know the true horrors of war—the deep psychological, physical, economical, and environmental damages and scars that result from one group's actions to prepare and wage violence against someone else. This *Reader* offers the most diverse set of voices ever collected together to try to describe those horrors, including essays written over many years about the often-unseen consequences of our actions. Read these stories in the hopes that future ones need never be written."

—PETER GLEICK
environmental scientist,
communicator

"*The War and Environment Reader* is packed with powerful writings, from classics by Margaret Mead to recent investigations of drone strikes, about why war happens, how it affects us, and how we can stop it. Everyone who cares about humanity's most urgent problem—in other words, everyone—should read this book."

—**JOHN HORGAN**
science journalist and
author, *The End of War*

"Anyone who studies environmentalism learns about climate change, biodiversity loss, and hazardous chemical pollution; but too often a major environmental threat is overlooked: war. Both caused by, and contributing to, environmental harm, war undermines our chance to live sustainably on this beautiful planet. Gar Smith's *War and Environment Reader* is a gift to our movement; it is a powerful collection of voices exploring war from many perspectives, and calling for a green and peaceful way forward. Please read this book and join the call!"

—**ANNIE LEONARD**
Executive Director, Greenpeace US;
author *The Story of Stuff*

"Even if war were moral, legal, defensive, beneficial to the spread of freedom, and inexpensive, we would be obliged to make abolishing it our top priority solely because of the case laid out so powerfully in this book: war is destroying this planet."

—DAVID SWANSON
Director, World Beyond War;
author, *War Is a Lie*

"This anthology is a much-needed volume of essays on the negative impact that the militarization of our world has on our planet—on the humans, animals, plants, oceans, lands, and air. The question is sometimes asked: 'what is war good for?' And the answer from millions of people is: 'Absolutely nothing!'"

—ANN WRIGHT
former colonel, U.S. Army; former diplomat;
antiwar activist

THE WAR AND ENVIRONMENT READER

THE WAR AND ENVIRONMENT READER

Gar Smith, *editor*

Just World Books is an imprint of Just World Publishing, LLC

Development editing: Brian Baughan
Project management and proofreading: Marissa Wold Uhrina
Typesetting: PerfecType, Nashville, TN
Cover design: *the*BookDesigners

Publisher's Cataloging-In-Publication Data
(Prepared by The Donohue Group, Inc.)

Names: Smith, Gar, editor.
Title: The war and environment reader / Gar Smith, editor.
Description: Charlottesville, Virginia : Just World Books, [2017]
Identifiers: LCCN 2017934406 | ISBN 978-1-68257-079-1 | ISBN 978-1-68257-080-7 (ePub) | ISBN 978-1-68257-081-4 (mobi) | ISBN 978-1-68257-082-1 (PDF)
Subjects: LCSH: War—Environmental aspects. | War—Economic aspects. | Military weapons—Environmental aspects. | Environmental degradation—Social aspects.
Classification: LCC TD195.W29 W37 2017 (print) | LCC TD195.W29 (ebook) | DDC 577.274—dc23

DEDICATION

To our despoiled and damaged world,
to the innocent victims of war,
to the thousands of activists killed
in the nonviolent struggle to defend Planet Earth;
to Dave Brower, who witnessed war and defended nature,
and to Doug Tompkins, who saw the need for this book.

Oilfields burning in Kuwait, 1991.

TABLE OF CONTENTS

Part II: Terracide—The War on Nature / 91

Nature in the Crosshairs

Collateral Damage

A Field Guide to Militarism

The Machinery of Mayhem

EarthTrust volunteers face the fiery aftermath of the 1991 Gulf War. William Thomas. Copyright 1991.

ENHEDUANNA'S LAMENT

One of the world's oldest surviving poems is an antiwar poem. It was written around 2300 BC in what is now southern Iraq by a Sumerian priestess named Enheduanna.

Lament to the Spirit of War

You hack everything down in battle. . . .
God of War, with your fierce wings
you slice away the land and charge
disguised as a raging storm,
growl as a roaring hurricane,
yell like a tempest yells,
thunder, rage, roar, and drum,
expel evil winds!
Your feet are filled with anxiety!
On your lyre of moans
I hear your loud dirge scream.
Like a fiery monster you fill the land with poison,
As thunder you growl over the earth,
trees and bushes collapse before you.
You are blood rushing down a mountain,
Spirit of hate, greed, and anger,
dominator of heaven and earth!
Your fire wafts over our land,
riding on a beast,
with indomitable commands,
you decide all fate.
You triumph over all our rites.
Who can explain why you go on so?
—Translation adapted by Daniela Gioseffi

A B-52 Stratofortress refuels before dropping bombs on Syria during Operation Inherent Resolve, 2017. PHOTO: SENIOR AIRMAN JORDAN CASTELAN.

INTRODUCTION: BEYOND PERMAWAR AND TERRACIDE

The environment has long been a silent casualty of war and armed conflict. From the contamination of land and the destruction of forests to the plunder of natural resources and the collapse of management systems, the environmental consequences of war are often widespread and devastating. Let us reaffirm our commitment to protect the environment from the impacts of war, and to prevent future conflicts over natural resources.

—UN secretary-general Ban Ki-moon, November 2014

Even before war breaks out, the Earth suffers. Minerals, chemicals, and fuels are violently wrested from Earth's forests, plains, and mountains. Much of this bounty is transformed into aircraft, gunboats, bullets, and bombs that further crater, sear, and poison the land, air, and water of our living planet. War is, and has always been, nature's nemesis.

In his 2001 book *War and Nature,* Edmund Russell observed:

> Since at least the days of the Old Testament, we have seen war and interactions with nature as separate, even opposite, endeavors. . . . Military historians have pushed beyond studies of battles and armies to examine the impact of military institutions on civilian society—but rarely on nature. Environmental historians have emphasized the role of nature in many events of our past—but rarely in war.

Environmental activists, however, have a surprisingly long history of confronting militarism. The environmental movement of the sixties emerged, in part, as a response to the horrors of the Vietnam war—Agent Orange, napalm, carpet-bombing—and the abiding threat of nuclear war.

In 1971, Greenpeace got its start challenging a planned U.S. nuclear test on one of Alaska's Aleutian Islands. The founders explained that they intentionally chose "Greenpeace" because it was "the best name we can think of to join the two great issues of our times, the survival of our environment and the peace of the world." Greenpeace's 1976 Declaration of Interdependence began: "We have arrived at a place in history where decisive action must be taken to avoid a general environmental disaster. With nuclear reactors proliferating and over 900 species on the endangered list, there can be no further delay or our children will be denied their future."

But challenging the profit-driven and the power-hungry—whether they are corporate polluters or the Pentagon itself—comes with risks. In June 1975, a team of Greenpeace activists attempted to disrupt the "Whale Wars" raging off the California coast by placing their rubber boat between a pod of whales and the guns of a Russian whaler. The activists were nearly killed when the Russians fired a harpoon directly over their heads.

On July 10, 1985, while I was visiting the Greenpeace International office in Amsterdam, we received some shocking news: the *Rainbow Warrior*, the organization's flagship, had been blown up in Auckland harbor and Fernando Pereira, a Greenpeace photographer, had been killed. It was later revealed that this well-planned, premeditated terrorist attack was an act of "state-sponsored terrorism." The French government had ordered a team of thirteen secret agents to sink the vessel and prevent Greenpeace from protesting a French atomic weapons test in the South Pacific.

Permawar

As the palace guard of the most expansive empire in world history, the Pentagon's operations impose unparalleled environmental impacts on the planet. The United States now maintains tens of thousands of troops stationed at more than 1,000 bases in more than 60 foreign countries. The Pentagon currently deploys troops in nearly 150 foreign lands. Entire generations have grown up in a culture that accepts the idea of "endless wars." In too many countries, culture, tradition, religion, government spin, and corporate propaganda conspire to glamorize the brandishing of weapons, the bursting of bombs, and the spilling of blood. It may not be good for war's victims and survivors, but it is provably beneficial to many people's "bottom line."

It is no surprise that war has become one of America's biggest exports. The grim truth is that war is "good for business." According to the venerable Swedish think tank the Stockholm International Peace Research Institute, in 2015 the world's top 100 arms manufacturers—excluding China, whose transactions were unreported—sold $307.7 billion worth of weapons, with U.S. corporations grabbing 56 percent of the sales.

Five of the world's largest war-profiteering companies are based in the United States—Lockheed Martin (the world's largest arms dealer, responsible for nuclear weapons, Trident submarines, Hellfire missiles), Boeing (B-52 bombers, JDAM smart bombs), Raytheon (Tomahawk long-range missiles, munitions), Northrop Grumman (Global Hawk drones, high-energy laser weapons), and General Dynamics (jet fighters, tanks, missiles, guns). In 2014, total sales for the top six U.S. arms makers topped $241 billion. Two-thirds of the world's arms makers are based in the United States or Western Europe, and they control more than 84 percent of the global arms trade.

Like any large, unregulated business, the misnamed "defense" industry routinely puts profits before people. In the past, the Pentagon made a habit of hosting peacetime military exercises to subsidize the arms industry. Now, the Pentagon and its contractors understand that "winning" wars is not as profitable as fighting unending "generational wars." Hence, the unwinnable "war on terror."

On September 10, 2001, Defense Secretary Donald Rumsfeld warned a large assembly of Pentagon employees of "an adversary that poses a serious threat to the security of the United States." That threat, he revealed, was "the Pentagon bureaucracy." "We cannot track $2.3 trillion in transactions," Rumsfeld confessed. (The Pentagon is the only federal agency exempt from a 1990 law requiring annual audits. In January 2002, *CBS News* reported that 25 percent of the Pentagon's annual spending was "unaccounted for."

In August 2016, the Pentagon's Inspector General raised the estimate of missing money to $6.5 trillion.)

Rumsfeld vowed "to wage an all-out campaign to shift [the] Pentagon's resources from bureaucracy to the battlefield." That never happened. On the day after his speech, Manhattan's Twin Towers and the Pentagon were attacked, and the defense industry was soon awash in new waves of unaudited dollars.

Terracide

Over the centuries, armies have fought for a wide array of reasons—land, water, food, slaves. Over the past sixty years, however, competition over natural resources has played an increasing role in driving armed conflict. According to a 2013 UN Environment Programme report, at least 40 percent of all intrastate conflicts involved the exploitation of natural resources—land, water, diamonds, timber, minerals, and, of course, oil.

In its FY 2010 *Base Structure Report*, the Pentagon disclosed that its global empire included more than 539,000 facilities at nearly 5,000 sites covering more than 28 million acres. With the world's largest air force and naval fleets, the Pentagon is the world's greatest institutional consumer of oil and a leading producer of climate-changing CO_2. The Pentagon has admitted to burning 320,000 barrels of oil a day—but that estimate ignores fuel consumed by its contractors and weapons producers.

Despite being the planet's greatest institutional emitter of climate-destabilizing gases, the Pentagon is exempt from reporting its pollution. This exemption covers weapons testing, military exercises, and "peacekeeping" missions. The United States insisted on this exemption during the 1998 Kyoto Protocol negotiations on the United Nations Framework Convention on Climate Change. After winning this concession, George W. Bush subsequently rejected the Kyoto Accord, claiming that it "would cause serious harm to the [oil-dependent] U.S. economy."

After the United States' disastrous 2003 invasion of Iraq, Gen. John Abizaid, former head of the U.S. Central Command, told a crowd at a 2008 Stanford University roundtable: "Of course, it's about oil; we can't really deny that." Defense Secretary Chuck Hagel, speaking with equal candor in

September 2007, told a group of Catholic University law students: "People say we're not fighting for oil. Of course we are."

Even when not engaged in outright acts of war, the Pentagon's far-flung operations impose damaging environmental burdens around the world. Inside America's borders, more than 11,000 military dumps simmer with a witches' brew of explosives, chemical warfare agents, toxic solvents, and heavy metals. The military is responsible for at least 900 of the country's 1,300 super-toxic "Superfund" sites.

Of course, the United States is not the only "superpower" to despoil the Earth. Russia's military has poisoned vast stretches of its own homeland—including boreal forests and Arctic waters. China's nuclear testing zone at Lop Nor and Britain's A-bomb range in Australia's Outback irradiated vast regions with cancer-causing fallout. But the scale of environmental destruction wrought by the U.S. military tops all the others.

Seeking Ecolibrium

One of the first mainstream environmental leaders to formally recognize warfare as an enemy of nature was David Ross Brower, who played a legendary role in the movement as executive director of the Sierra Club and founder of the Friends of the Earth (FOE) and Earth Island Institute. In 1970, Brower and other environmental leaders sent a telegram to President Richard Nixon protesting the devastation of the U.S. war in Vietnam. Brower followed up with an "open letter" from FOE. The statement, titled "Ecology and War," was designed to run as a full-page ad in the *New York Times*. (Twelve years later, Brower included the statement as a centerfold ad in the October 1982 issue of FOE's newsmagazine, *Not Man Apart*. This time the message was addressed to President Ronald Reagan.)

"Until recently," Brower wrote, "we were content to work for our usual constituency: Life in its miraculous diversity of forms. . . . We have left it for others to argue about war."

That had to change, he explained, because "countries obsessed with the glamours of militarism remain blind to other ecological perils." Citing a 1980 White House report, Brower ticked off a host of impending environmental threats—global water shortages, loss of forest cover, expanding deserts, increasing levels of climate-altering gases, the destruction of wildlife habitats—all signs of ecosystems in collapse or, as Brower's ad put it, "a Holocaust, except in slow motion, without the fireball."

His appeal ended with a call for "a new political language" that acknowledged the futility of "belligerency [and] competition for resources on a finite planet." He argued that "exploitation of people and nature are behaviors that have failed. They are out of date. We need to seek people skilled in the arts of peace. . . . We need to attend to other problems—poverty and the coming breakdown of the planet's life-support systems."

Brower's plea was taken up more than two decades later when the 1992 United Nations Conference on Environment and Development issued the Rio Declaration, which stated: "Warfare is inherently destructive of sustainable development. States shall therefore respect international law providing protection for the environment in times of armed conflict."

This was hardly a novel observation. In his classic fifth-century BC study, *The Art of War,* Chinese military strategist Sun Tzu wrote: "No country has ever profited from protracted warfare." Instead of promoting scorched-earth tactics, the wise general urged soldiers to "fight under Heaven with the paramount aim of 'preservation.'"

Even the Pentagon has appeared increasingly ready to recognize the strategic value of maintaining a sustainable planet. In October 2014, a Pentagon report identified climate change as an "immediate threat" to national security, echoing a 2003 Pentagon prediction that climate change could "prove a greater risk to the world than terrorism." While the *2014 Global Terrorism Index* listed 100,000 lives lost to terrorist acts over a period of 13 years, the 2010 edition of the *Climate Vulnerability Monitor* reported that climate change claimed 400,000 lives and cost the world economy $1.2 trillion—in a single year.

In June 2016, *National Geographic* (citing research findings from Amsterdam's Vrije Universiteit) warned that new perils were likely to arise from the collision of war and nature. "Wars, murders, and other acts of violence will likely become more commonplace in coming decades as the effects of global warming cause temperatures to flare worldwide." An interdisciplinary team of researchers from Princeton and the University of California at Berkeley have predicted incidents of personal violence, civil unrest, and war could increase 56 percent by 2050 as the planet warms, accelerating droughts, floods, disease, crop failures, and mass migrations.

Our Troubled Path

Today, we stand at a precipice. Biospheres are collapsing, a Sixth Extinction is underway, and, on January 2017, the Bulletin of Atomic Scientists' "Doomsday Clock," which illustrates how close the planet stands to a nuclear holocaust, was moved to 2.5 minutes before midnight. That chilling recalibration was attributed to the global rise of "strident nationalism" and the bellicose posturing of Donald J. Trump—an impetuous leader who denies the reality of climate change and seems oblivious to the risks of thermonuclear war.

As a presidential candidate in 2016, Trump told MSNBC's Chris Matthews: "Somebody hits us within ISIS—you wouldn't fight back with a nuke?" When Matthews protested that "nobody wants to hear that" from someone running for president, Trump shot back: "Then why are we making them?" Trump has repeatedly refused to take the nuclear option "off the table," explaining to *Meet the Press* host John Dickerson in 2016 that, as a world leader, "You want to be unpredictable."

In one of his early acts as president, Trump endorsed President Barack Obama's unconscionable plan to spend $1 trillion rebuilding America's "nuclear arsenal." Trump had long made his reputation and his fortune by acting as a crude bully. But the world has too many bullies spoiling for a fight.

In his 1949 book, *The Immense Journey: An Imaginative Naturalist Explores the Mysteries of Man and Nature*, the American anthropologist, educator, and natural science writer Loren Eiseley offered a different vision: "The need is now for a gentler, a more tolerant people than those who won for us against the ice, the tiger and the bear. The hand that hefted the ax, out of some old blind allegiance to the past, fondles the machine gun as lovingly. It is a habit we will have to break to survive, but the roots go very deep."

In 2017, polluter-versus-protector tensions escalated dramatically when Donald Trump marched into the White House. Trump's inaugural budget called for defunding a range of humanitarian programs—health and human services, education, arts, Meals on Wheels, and environmental protections—to shift another $54 billion to the Pentagon. Following through on his campaign declaration to "bomb the shit out of [ISIS]! I don't care," Trump increased airstrikes in Afghanistan, Iraq, Syria, and Yemen, resulting in the deaths of hundreds of civilian men, women, and children—and triggering a new spike in anti-American anger.

On the home front, Trump revived controversial oil pipeline projects, renewed support for coal mining, gutted anti-pollution regulations, and targeted Obama's Clean Energy Plan. The Environmental Protection Agency (EPA) was singled out for special abuse. Under Scott Pruitt, a former Oklahoma attorney general who had threatened to destroy the agency he now headed, the EPA's funding was to be slashed by 31 percent. More than 3,000 employees were fired and the surviving staffers were reportedly banned from using the phrase "climate change."

Protesting a so-called "War on Coal," Trump reasserted the industry's "right" to dump mine tailings into mountain streams. And, on June 1, 2017, he announced his rejection of the Paris Climate Accord. He shredded pollution reduction goals for U.S. automakers, declared his intent to open public lands to oil and gas drilling, and filled his cabinet with a swamp-load of wealthy, carbon-friendly, pro-business and big-bank insiders.

Trump also brought unique conflicts of interest to his role as commander-in-chief. When he launched his unconstitutional attack on Syria's al-Shayrat airfield on April 7, 2017, Trump may have been inflating not only his reputation as a powerful world leader but his personal fortunes as well. Unlike previous U.S. presidents, Trump holds investment properties around the world and owns shares in hundreds of companies. As recently as 2016, Trump's investments included Raytheon, maker of the fifty-nine Tomahawk cruise missiles fired at Syria. When Trump launched $93.8 million worth of missiles that week, the value of Raytheon stocks soared.

While Trump's tax forms remain a state secret, *Fortune* magazine reported in August 2016 that his top holdings include shares in Phillips 66, a major oil company. Trump's 2016 federal disclosure forms show he owned stock in Energy Transfer Partners, the firm behind the Dakota

Access oil pipeline. President Obama halted the construction of the pipeline on environmental grounds; Trump revived it.

Unity over Division

The rich and powerful clearly understand the strategy of "divide and conquer." The "one percent" has less to fear from a "99 percent" subdivided into scores of competing factions. Civil discord, when it turns violent, is the preamble to civil war. Instead of working together to find cures to existential problems—rising temperatures, sea levels, and populations; falling incomes, water tables, and food harvests—we could find ourselves immobilized by suspicion, anger, and fear.

In her 2017 Earth Day statement, U.S. Representative Tulsi Gabbard (D-HI) declared: "Change is coming and it's up to us whether that change will be transformative or catastrophic. We don't have time to fight amongst each other. We don't have time to engage in partisan rhetoric and bickering. We must be the change we want to see in the world. The future of the human race depends on it."

Instead of fortifying our cities and neighborhoods against one another, we need to forge coalitions to promote common-ground solutions that will protect our shared, life-sustaining planetary biosphere. Instead of promoting competition and "fighting to win," we need to exercise compassion and learn how to cooperate to survive.

Our hope is that this collection of essays will illuminate the economic, political, and cultural drivers of permawar, detail the cascading dangers of terracide, and provide guideposts and goals for achieving ecolibrium—a human presence on Earth that is more in balance with nature. One way or another, our planet will be transformed. Let's strive for a peaceful, just, and sustainable future.

Part I

PERMAWAR—HUMAN NATURE *and* WARFARE

I knew Man was doomed when I realized that his strongest inclination was toward ever-increasing homogeneity—which goes completely against Nature. Nature moves toward ever-increasing diversity. Diversity is Nature's strength. Nature loves diversity.

—Yvon Chouinard, founder of the
outdoor-gear company Patagonia

In the Hebrew Bible, in Genesis 6:7, the first "war on nature" is launched by an angry God who destroys his Creation—animals, birds, insects, fish, forests, mountains, and meadows—because of the "wickedness" of humans. But God came to regret the destruction, as recorded in Genesis 8:21–22: "I will never again curse the ground because of man. . . . Neither will I ever again strike down every living creature as I have done. For all the days of the Earth, seedtime and harvest, cold and heat, summer and winter, day and night, shall not cease."

Unfortunately, humankind failed to make a similar pledge.

Permawar—Human Nature and War explores the interaction between human nature and the war on nature by asking, "Are there innate forces in the human spirit—or social and cultural influences—that predispose individuals and groups to choose killing over cooperation?"

In a darkly prescient 2006 essay posted on *OpEdNews*, University of Texas journalism professor Robert Jensen raised an intriguing question: "Can a nation have a coherent character?" When he searched the American Psychiatric Association's *Diagnostic and Statistical Manual* for a clue to "America's national character," one category jumped out: "Narcissistic Personality Disorder." NPD's signature traits include "a pervasive pattern of grandiosity (in fantasy or behavior), need for admiration, and lack of empathy." So perhaps it is no surprise that, ten years after Jensen's diagnosis, Donald Trump bellowed his way into the Oval Office, boastful and belligerent, and set about gutting environmental protections and financial regulations to unleash the forces of carbon-fueled capitalism.

According to Politifact, the United States spends more on its military than the next eight militarized countries combined. Yet, despite massive increases in military spending, the United States has not won a war since 1945, nor has it managed to bring democracy or freedom to any of the many nations it has attacked or occupied over the past seven decades. Instead of drawing useful conclusions from this sorry history, Republicans and Democrats alike have asked Americans to accept a future of "generational" wars. "Disaster capitalism" has turned combat into a form of commerce. Wars are no longer avoided, they are provoked—often through fictitious "false flag" incidents designed to mislead and stampede the public. Meanwhile, a small, powerful elite reaps massive dividends by investing in armaments—including nuclear weapons.

In a January 2017 report titled *An Economy for the 99%*, Oxfam revealed that eight super-rich men controlled $438 billion—as much wealth as 3.6 billion of the world's poorest people. At the same time, the Pentagon's FY 2016 budget was $521.7 billion. Redistributing that wealth could make every homeless person in the United States a millionaire, concluded Adrienne Mahsa Varkiani of ThinkProgress in a May 2016 article. The $400 billion tab for Lockheed Martin's underperforming F-35 Joint Strike Fighter could have allowed the National School Lunch Program to feed 31 million American children—for 24 years. And yet, with one-fifth of America's children malnourished, both political parties continue to prioritize the feeding of Pentagon contractors—including $1 trillion to create "a new generation" of nuclear bombs over the next thirty years.

Mapping the Terrain

The Roots of War examines foundational forces that underlie the militaristic mindset—including patriarchy, machismo, and misogyny. Other factors include: linear thinking that divorces the human mind from nature's rhythms; a celebration of competition and combative sports; and a torrent of Hollywood blockbusters that promote fists and firearms as problem-solving tools. And whenever force rules the day, Nature takes the hits.

As the United Nations Environment Programme has noted over the past sixty years, nearly half of all internal conflicts involved battles for control of natural resources. Millions of people have perished in bloody skirmishes over diamonds, timber, oil, and gold. Now, as global temperatures and tensions rise, drought, floods, hurricanes, and disease are driving new "resource wars."

The Business of War examines militarism as an economic force. As America's most decorated Marine once revealed, "War is a racket"—a system that is ruinously expensive to the people who shoulder the costs but incredibly remunerative to the hidden few who reap incredible fortunes despite never having to shoulder a rifle.

With the global economy staggering under burdens of unpayable debt, war has become an economic stimulus program. Despite a towering federal debt topping $14 trillion (according to a September 2016 Congressional Budget Office estimate), Donald Trump, in his first week in office, called on Washington to lavish even more money on "our depleted military." (A September 2016 study by Brown University's Watson Institute predicted interest on the US debt could top $1 trillion by 2023 and exceed $7.9 trillion by 2053.)

Meanwhile, the War Economy keeps its eye on the future—cultivating the next generation of soldiers by putting toy guns and violent videogames into the hands of children.

The Aftermath of War tracks outcomes that can be grim and long lasting. Ten percent of the 2.7 million tons of Allied bombs dropped on Europe during World War II failed to detonate and continue to threaten modern towns and villages. Blasted metal skeletons of ships and planes litter bays and beaches across the South Pacific. Exposure to toxic burn pits, radioactive debris, and chemical agents have claimed the lives of thousands of military men and women while civilians fight to survive the cancers caused by fallout carried downwind from America's atomic playground at the Nevada Test Site. On South Pacific islands exposed to fallout from U.S. nuclear tests, horrified mothers gave birth to "jellyfish babies." In the crowded hospitals of Iraq, women from Fallujah and other cities blasted by U.S. depleted uranium weapons continue to deliver deformed babies—children born with the wrong number of fingers, legs, arms, eyes, and heads. Combat veterans struggle with PTSD, trying to block the grotesque memories of violent deaths while considering the relief of suicide.

Fortunately, nature is resilient—despite wounds to the land that can outlast decades. One of the best examples of nature's ability to heal is found in the Demilitarized Zone between the two Koreas. Since 1953, landmines, barricades, and armed sentries have prevented human encroachment inside the DMZ's 400 square miles. Protected from human presence, the DMZ has become a "new Eden," one of the most flourishing and biodiverse wilderness areas on the planet.

The success of this Korean no-man's-land brings to mind Rainer Maria Rilke's wistful observation: "If we surrendered to Earth's intelligence, we could rise up rooted, like trees."

THE ROOTS OF WAR

Patriarchy both creates the rage in boys and then contains it for later use, making it a resource to exploit later on as boys become men.

As a national product, this rage can be garnered to further imperialism, hatred, and oppression of women and men globally. This rage is needed if boys are to become men willing to travel around the world to fight wars.

—bell hooks, American author,
feminist, and social activist

Stones to Drones: A History of War on Earth

Gar Smith, Environmentalists Against War (2017)

From the biblical battle of Schechem to the proxy wars of the twenty-first century, human warfare has despoiled the land—ravaging crops, flattening forests, and spilling shrapnel, poisons, and blood over the Earth. Warfare has scorched and scarred the face of the natural world with wounds that have outlasted the reign of warrior kings and legions of common soldiers.

About 14,000 years ago, North American hunters developed the "Clovis spearpoint," a deadly tool that gave our ancestors the ability to kill large predators, including mastodons and saber-tooth tigers. Hominids suddenly vaulted to the top of the food chain—a major step on the path to dominating the natural world.

But there is scant archaeological evidence that organized warfare existed among humans before 4,000 BC. Anthropological findings suggest that warfare is a social invention that first appeared around 6,000 years ago with the development of centralized states, patriarchy, and slavery. For most of the preceding 100,000 years, human history was free of large-scale violence.

In January 2016, however, this consensus was rocked by a grim discovery in the hard clay of the Nataruk archaeological site near Kenya's Lake Turkana. The journal *Nature* reported that the buried bones of 27 men, women, and children killed 10,000 years ago bore clear evidence of a violent ambush. Ten of the victims died from crushing blows to the head and stab wounds. A stone arrowhead was still embedded in the skull of one victim. This finding marks the first evidence of prehistoric "war-like" violence among a hunter-gatherer community. Cambridge University researchers speculated that the victims may have developed the ability to fashion blades and pottery, giving rise to the possession of property that could have prompted the brutal attack.

Still, evolutionary science largely suggests that humans are not predisposed to violence: human behavior is marked by both aggression and compassion. Modern warfare is a highly organized form of learned behavior imposed to support "warrior cultures" (aka "dominator societies") where men and boys are encouraged to dominate other males, to subjugate women, to compete in physical contests, and to view nature as a prize to be conquered and exploited.

Until quite recently, human combat was limited to the range of a clenched fist, the sweep of a sword, or the arc of a thrown spear. However, as the tools of deadly force continued to evolve, stones and sticks were supplanted by lances and maces, swords and trebuchets, machine guns and cluster bombs. The bow and arrow made it possible to kill at a distance. The invention of gunpowder further expanded the range of disembodied combat. The "honorable" convention of a two-man duel gave way to the mass slaughter of modern warfare with uniformed battalions ordered to run screaming into curtains of gunfire.

Ancient Tales of Blood and Conquest

The Epic of Gilgamesh, one of the world's oldest tales, recounts a Mesopotamian warrior's quest to kill Humbaba—a monster who reigns over a sacred Cedar Forest—and seize the prized trees as plunder. But Humbaba was not only the protector of cedars, he was also the servant of Enlil, the god of earth, wind, and air. By killing this protector of Nature, Gilgamesh called down a curse upon his own head.

The Bible offers further tales of environmental war. Judges 15:4–5 relates the story of Samson's unorthodox plan to attack the Philistines: "[H]e went out and caught three hundred foxes and tied them tail to tail in pairs. He then fastened a torch to every pair of tails, lit the torches, and let the foxes loose in the standing grain of the Philistines. He burned up the shocks and standing grain, together with the vineyards and olive groves."

Judges 9:45 records how Abimelech conquered the city of Schechem and, after killing its people, "destroyed the city and scattered salt over it." The salt-sowing tactic (an early use of chemical warfare) was famously employed during the Third Punic War, when Roman invaders salted the land around Carthage, leaving the soil infertile.

In 429 BC, during the Peloponnesian War, King Archidamus began his attack on Plataea by felling the fruit trees surrounding the fortified town. His attempt to destroy the city by launching bundles of pitch-and-sulfur-soaked wood over the walls was thwarted when nature intervened with a rainstorm that extinguished the flames.

Sometimes, the environmental damage from war is self-inflicted. In 480 BC, Pericles directed retreating Athenians to destroy their homes and land as they fled the Persians. During Genghis Khan's advance through Asia and Eastern Europe in the thirteenth century, livestock and crops were preemptively destroyed lest they fall into the hands of the invading Mongols. When Genghis Khan reached Baghdad, one of his first targets was the ancient system of waterworks along the Tigris River—an engineering marvel that had provided the city with clean water for 2,000 years.

In 1346, Mongol Tartars laid siege to Caffa, a port city on the Black Sea. Caffa became the target of the earliest recorded incident of biological warfare when the attackers catapulted bodies of plague victims over the fortified walls. (The few survivors who escaped took the "Black Death" with them when they fled to Italy.)

Poisoning water supplies and destroying crops and livestock has always been an effective means of subduing a population. These "scorched-earth" strategies remain a preferred way of dealing with agrarian societies—as we have witnessed in numerous campaigns against people in the Global South, from Nicaragua and Haiti to the Philippines and Vietnam.

America's Indian Wars

During the American Revolution, Gen. George Washington ordered Gen. John Sullivan to direct "scorched-earth" tactics against Native Americans who had allied themselves with British troops. The fruit orchards and corn crops of the Iroquois Nation were razed in hopes that their destruction would cause the Iroquois to perish over the harsh winter.

A century later, during the U.S. Army's Indian Wars in Arizona, the Navajo peoples' corn, orchards, sheep, and other livestock were destroyed by fire, ax, and buckshot. In 1869, Civil War general William Tecumseh Sherman advised President Ulysses S. Grant: "We must act with vindictive earnestness against the Sioux, even to their extermination, men, women, and children."

In 1865, Gen. Philip Sheridan launched a genocidal war against the Plains Indians that involved the mass slaughter of the native bison that once numbered near 60 million. Ghastly photos from that era document the effectiveness of the Buffalo Holocaust. The black-and-white images record mile after long mile of dark hides stacked 8 to 10 feet high. By 1893, fewer than 400 bison remained. As Chief Plenty Coups of the Crow Nation lamented: "When the buffalo went away, the hearts of my people fell to the ground."

The call to conquer both nature and Native people was given full voice by the great American poet Walt Whitman who, in 1865, provided this anthem for Western expansion:

Come, my tan-faced children,
Follow well in order,
Get your weapons ready;
Have you your pistols?
Have you your sharp-edged axes?
Pioneers! O Pioneers!

The American Civil War

Gen. Sherman's "March through Georgia" and Gen. Sheridan's campaign in Virginia's Shenandoah Valley were examples of "total war"—a scorched-earth approach aimed at destroying civilian crops, livestock, and property. Sherman's army devastated an estimated 10 million acres of land in Georgia while causing an estimated $4 million in losses to Mississippi's natural resources ($75 million in 2016 dollars). When then-general Ulysses S. Grant ordered Sheridan to turn the Shenandoah into a barren wasteland, Sheridan promised there would be little left "for man or beast." Shenandoah's farmlands were transformed into fire-blackened landscapes that shocked and demoralized the civilian population.

Ambrose Bierce vividly described the environmental impacts of the Civil War: "Riven and torn with cannon-shot, the trunks of the trees protruded bunches of splinters like hands, the fingers above the wound interlacing with those below. The bark of these trees, from the root upward to a height of 10 or 20 feet, was so thickly pierced with bullets and grape that one could not have laid a hand on it without covering several punctures."

Fire was a particularly devastating by-product of war as fallen leaves, severed tree limbs, and wooden breastworks spread the flames. Massive entrenchments also left an enduring mark. The scars of battlements and trenches dug into the earth at North Carolina's Bentonville Battlefield are still visible more than 150 years later.

War in a New Century

During the many horrors of World War I, the greatest environmental impacts were seen in the fields of France. The Battle of the Somme (where 57,000 British soldiers died on the first day of battle) left the countryside disfigured by trenches that are evident a century later. In a single day, Thiepval Ridge was stripped of vegetation, and the once-verdant High Wood was left a burnt shambles of blasted, mangled trunks. Nearly 250,000 acres of farmland were ruined to the point that they had to be abandoned.

The battle zone encompassed 1.5 million acres of French forests and destroyed an estimated 494,000 acres. (The forests had not yet recovered when World War II erupted twenty-one years later.) In Poland, German troops leveled forests to provide timber for military construction and, in the process, destroyed the habitat of the endangered wisent (European buffalo). Deprived of shelter, the small population of surviving wisents was quickly cut down by the rifles of hungry German soldiers.

In 1918, a German officer who survived "the Great War" described the battlefield as a landscape of "dumb, black stumps of shattered trees which still stick up where there used to be villages. Flayed by splinters of bursting shells, they stand like corpses upright. Not a blade of grass anywhere. Just miles of flat, empty, broken, and tumbled stone." World War I's bloody battles claimed the lives of soldiers, civilians, domestic animals, and wildlife. A century after the carnage, Belgian farmers were still unearthing the bones of soldiers who bled to death in Flanders Fields.

During World War I, Great Britain enlisted 571,000 pack animals to move its men and weapons (more than 68,000 horses and mules were killed in that war). It took forty pounds of fodder and eight gallons of water to support each horse for one day. Not surprisingly, British ships wound up hauling 5.4 million tons of hay and oats to France—outweighing shipments of ammunition.

World War I dramatically demonstrated how fighting wars abroad can have serious repercussions at home. In her 1939 book, *America Begins Again*, Katherine Glover revealed how World War I inflicted historically unprecedented damage on the U.S. environment when the struggle in the battlefields of Europe "spread to the cotton fields of the South, the cornfields of the Middle West, and the wheat fields of the Great Plains." In order to feed the war effort, 40 million acres of land were rushed into cultivation. But this acreage was largely unsuited for agriculture. Across the Northwest, lakes, reservoirs, and wetlands were drained to create farmland. In the Southwest, native grasses were uprooted to make space for wheat fields. In Minnesota, Wisconsin, and Michigan, entire forests were clear-cut to serve wartime needs. In the South, extensive overplanting of cotton led to depleted soils that eventually succumbed to drought and erosion.

The oil-fueled mechanization of war further transformed—and expanded—the battlefield. Instead of relying on horses and mules, modern armies now rely on trucks, armored tanks, and aircraft. The U.S. Army's first use of a truck occurred during a 1916 invasion of Mexico in pursuit of Pancho Villa. By the end of World War I, the General Motors Corporation had built 8,512 military vehicles and earned a tidy profit. Modern mechanized armies are no longer dependent on oats and hay. The "blood" that fuels the world's ships, tanks, choppers, and bombers is petroleum.

Air power was another historic game-changer because it enabled generals and admirals to order bombs to be dropped at great distances and permitted pilots to search the terrain for signs of the "enemy" from a platform of relative safety. But because woodlands and forests sometimes provide cover for insurgents on the ground, forests also became military targets requiring the "scorched earth" treatment—whether by bomb blast, napalm, or chemical spray.

World War II

With the outbreak of World War II, the European countryside suffered a renewed onslaught. As the German army marched into The Netherlands, the Dutch population blocked its advance by opening dikes and irrigation gates to flood the country's roads with seawater. (In 1672, the Dutch used the same ploy to quench the invasion plans of French king Louis XIV.) Holland's lowlands suffered even worse damage after German troops captured the floodgates and attempted to starve the Dutch resistance by covering 17 percent of the country's farmland with saltwater.

The Germans weren't alone in using water as a weapon of war. In May 1943, Allied bombers took out two large dams in Germany's Ruhr Valley. An estimated 34.3 billion gallons poured from the Möhne Dam, and another 52.8 billion gallons burst from the Eder Dam. Nearly 7,500 acres of German farmland were submerged, along with more than 6,500 cattle, pigs, and other livestock.

The Germans occupied Norway from 1942 to 1945, but when they began to lose ground to the advancing Allies, Hitler's retreating troops set about destroying everything of potential value. They devastated 15 million acres, methodically eradicating buildings, roads, crops, forests, water supplies, and wildlife. Fifty percent of Norway's 95,000 reindeer were killed.

During Germany's occupation of France, nearly one million acres of forest were laid waste by bombs, shells, and bullets. As many as 247,000 acres of forest went up in flames while bomb blasts and shrapnel stripped vegetation from slopes, triggering erosion on a massive scale.

Fifty years after the end of World War II, millions of bombs, artillery shells, and underwater mines had been recovered from the fields and waterways of France. A century later, millions of acres remain off-limits and the "Iron Harvest" of buried ordnance still claims an occasional victim.

Wars in the Pacific

Many of the South Pacific's islands—isolated outposts of untouched wilderness with delicately balanced ecosystems—suffered mightily during World War II. During the Battle of Corregidor, a dense and thriving jungle was blasted into a graveyard of bodies, seared and broken trees, obscured by dust storms and palls of acrid smoke rising from smoldering palms and grassfires.

The Pentagon employed the same approach for each island attacked. First, a naval blockade followed by artillery shelling. Next, extended bombing by aircraft. Finally, troops would be ordered ashore. Landing craft would sweep in over coral reefs to secure the beachhead. U.S. soldiers would torch suspected hideouts with flamethrowers as bulldozers cleared the land.

To make way for the construction of airfields, topsoil and trees were removed. The most conveniently available material for building runways was coral—preferably live coral harvested from beneath the waves and crushed. On some South Pacific islands, the coral never recovered.

South Pacific wildlife took a major hit during World War II, with nesting places destroyed, migration patterns interrupted, birds killed, and eggs smashed by the thousands. Many species were eradicated during the Pacific War—mainly through habitat loss. An unknown number of whales were killed by naval gunners who mistook the cetaceans for enemy subs. Oil spills, following the bombing and sinking of supply tankers, caused uncounted losses of marine life. It is estimated that 300 tankers exploded, ruptured, spilled, or sank during the World War II.

World War II's most destructive event took place with the detonation of two nuclear bombs over the Japanese cities of Hiroshima and Nagasaki in August 1945. The fireball and blast was followed by a "black rain" that pelted the survivors for days, peeling away the scorched soil and leaving behind an invisible mist of radiation that seeped into the water that people sipped and infiltrated the air they breathed. This ghostly pall of radiation left a ghastly legacy—cancer spikes and mutations in plants, animals, and children.

Wars in the Middle East and Africa

During its 2003 military invasion of Iraq, the United States rained more than 1,000 tons of depleted uranium over the land. These weapons left a legacy of cancers, stillborn babies, and a generation of horrifically deformed children in Fallujah and other cities—a devastating blow for a country that also lost a million children to starvation and disease caused by U.S.-imposed sanctions.

In 1991, an apocalyptic shroud of smoke from Kuwait's burning oil fields turned day to night and released vast plumes of contaminated soot that blackened snows in the Himalayas. In 1999, NATO's intervention in Yugoslavia led to the bombing of a petrochemical plant in Panĉevo, which sent clouds of toxic chemicals into the air and spilled tons of pollution into nearby rivers.

From 1992 to 2007, U.S. bombing and illegal logging (committed by desperate Afghan refugees and U.S.-backed warlords) destroyed 38 percent of Afghanistan's forest habitat, triggering an 85 percent drop in the number of birds visiting the region.

In Africa, the Rwandan war drove nearly 750,000 people into the forests of Virunga National Park. Desperate refugees were soon clearing around 1,000 tons of wood from the park—every day. By the time the war ended, Worldwatch Institute reported, 105 square miles had been ransacked and 35 square miles were "stripped bare."

In Sudan, desperate soldiers and civilians spilled into Garamba National Park, decimating the animal population. In the Democratic Republic of the Congo, armed conflict caused the resident elephant population to plummet from 22,000 to 5,000. In 2006, Mai-Mai rebel fighters in the DRC killed every last hippopotamus occupying the country's two major rivers in just two months.

The Radioactive Legacy of the Cold War

The postwar decades gave rise to a quiet escalation of environmental damage as the Cold War unleashed an unchecked competition for newer and deadlier weapons. Foremost among them, nuclear bombs. Before the Nuclear Test Ban Treaty was signed in 1963, 1,352 underground nuclear blasts, 520 atmospheric detonations, and eight sub-sea explosions had jolted the planet and its oceans. The blasts totaled 535 megatons—equal to 36,400 Hiroshima-sized detonations.

In 2002, the National Cancer Institute (NCI) and the National Centers for Disease Control reported that everyone on Earth had been exposed to radiation released by open-air nuclear testing. According to the NCI's 2002 estimate, nuclear fallout killed more than 15,000 Americans and caused "at least 80,000 cancers."

The impact is on a scale with global warming, but the signs of radioactive "global harming" are subtle and cellular—less visible than droughts, firestorms, hurricanes, and rising seas. This lethal legacy of the Cold War remains a persistent and personal reminder that war's impacts on the environment profoundly inhabit the fate of every living creature on Earth today.

Drones and Warbots: The Future of War

The invention and deployment of armed drones as weapons of war marked a new achievement in the annals of technology and cowardice. Armed with devastating air-to-ground missiles, U.S. Predator and Reaper drones began to patrol African and Middle Eastern skies guided by "warriors" sitting in front of computer screens, twiddling joysticks in air-conditioned "arcades" safely ensconced in the Nevada desert, thirty-five miles from Las Vegas.

The United States first deployed its fleet of winged robots during the 1994 Balkans War. The CIA conducted its first "drone assassination" on February 4, 2001, when, in an attempt to kill Osama bin Laden, a drone opened fire on a gathering in Afghanistan's Paktia Province. The CIA acted after identifying "a tall man" in the crowd. The victims turned out to be a group of innocent civilians out collecting scrap metal.

Drone attacks and "signature strikes" (assassination attempts personally authorized by the president) have killed hundreds of innocent civilians in Pakistan, Afghanistan, Yemen, and Somalia—and driven thousands of angry survivors into the ranks of anti-U.S. militias. The civilian toll has been alarming. According to the Bureau of Investigative Reporting, between January 2005 and April 2017, 2,215 "confirmed" U.S. drone strikes were responsible for 6,206 to 8,906 deaths—including 736 to 1,391 civilians and 242 to 307 children.

In addition to robots in the skies, the Pentagon also is preparing to deploy robot boots on the ground. In 2007, the Defense Advanced Research Projects Agency (DARPA) field-tested a new Goliath-like weapon in Iraq. The Special Weapons Observation Remote Direct Action System (SWORDS) was a ground-based, radio-controlled killer droid fitted with video eyeballs, tank treads, and an M249 light machine gun. Three SWORDS units were deployed to Iraq in 2007, but they failed to meet expectations. According to Foster-Miller, the manufacturer, the SWORDS units suffered "some technical issues" and, as of this writing, the program has not been refunded.

While a Predator drone is about the size of a school bus, the newest members of the Pentagon's arsenal include small-scale "war-bots." Some are designed to look like and fly like birds while others crawl like insects. Sandia National Laboratory has built a remote-controlled fly-bot weighing one ounce and smaller than a dime. This mini-bot comes equipped with a TV scanner, microphone, and a chemical microsensor. Spy-bots, another form of micro-weaponry, can be released in swarms to covertly infiltrate entire neighborhoods. They can follow targets down streets and slip through open windows and doors. Some can even quietly hover behind a victim's head before triggering a lethal explosion.

Boston Dynamics has created the Big Dog—a four-legged, gas-powered metallic horse that can navigate mountainsides and frozen lakes while carrying 400 pounds of weapons and supplies (four times more than a human soldier can carry). Meanwhile, a UC Berkeley research team has created a "force multiplier" called the Human Universal Load Carrier (aka "the HULC"), a mechanical exoskeleton that can be strapped to the human body to endow the wearer with superhuman strength.

In South Korea, Samsung robots have replaced human guards along the border with North Korea. When these robots detect the approach of a human, they demand a password. If they don't hear the correct answer, they automatically open fire.

DARPA's ultimate goal is to "take the 'man' out of 'unmanned' warfare" by building war-bots with the capacity to make their own decisions about who to kill and when. A U.S. Joint Forces Command study predicts deployment of fully autonomous battlefield robots by 2025. These super-strong, super-smart war-bots will inevitably make mistakes and kill innocent civilians but, unlike humans, they won't experience guilt or suffer PTSD.

On November 21, 2012, the Pentagon issued a fifteen-page directive describing an autonomous weapons system that "once activated, can select and engage targets" without human supervision. The Pentagon has envisioned a $130 billion Robot Army. The Future Combat Systems program—one of the largest military contracts on record—would replace human soldiers with mechanical Terminators that could be deployed at one-tenth the cost of a flesh-and-bones warrior. Eliminating humans from the combat workforce would save billions in combat pay and veterans benefits.

There is resistance. In 2009, the International Committee for Robot Arms Control announced the launch of the Campaign to Stop Killer Robots. In May 2013, a UN report called for a ban to prevent a robot arms race, noting that "tireless war machines, ready for deployment at the push of a button, pose the danger of permanent armed conflict."

From Conquest to Stewardship

From the War of Independence through 2006, U.S. Navy historians have recorded 234 cases in which the United States has used its armed forces abroad "in situations of conflict or potential conflict or for other than normal peacetime operations." Between 1945 and 2014, the United States was responsible for launching 201 out of 248 significant world conflicts. Since the Pentagon's retreat from Vietnam in 1973, the United States has attacked, invaded, waged covert wars against, or attempted to overthrow governments in Afghanistan, Angola, Argentina, Bosnia, Cambodia, El Salvador, Grenada, Haiti, Iran, Iraq, Kosovo, Kuwait, Lebanon, Libya, Nicaragua, Pakistan, Panama, the Philippines, Somalia, Sudan, Syria, Ukraine, Yemen, and the former Yugoslavia. None of these countries were transformed into democracies as a result of U.S. intervention.

In early 2017, instead of abolishing national stockpiles of nuclear weapons, President Donald Trump called for moving funds from disarmament and military cleanup programs to fund the renovation of America's nuclear arsenal. (Trump was not breaking with tradition. In August 2015, on the seventieth anniversary of the U.S. bombing of Hiroshima and Nagasaki, Nobel Peace Prize laureate Barack Obama called for spending nearly $1 trillion on a new generation of "modernized" nuclear weapons.)

In a June 2015 essay for *The American Conservative*, Jon Bosil Utley noted that America no longer actually aims to "win" its wars, "because winning a war is secondary to other goals in our war-making. Winning or losing has little immediate consequence for the United States because the wars we start—wars of choice—are not of vital national interest; losing doesn't mean getting invaded or our cities being destroyed."

As former congressperson Ron Paul has pointed out: "Every dollar or euro spent on a contrived threat is a dollar or euro taken out of the real economy and wasted on military Keynesianism. Such spending benefits a thin layer of well-connected and well-paid elites. It diverts scarce resources from meeting the needs and desires of the population and channels them into manufacturing tools of destruction." It's not war that frightens these corporations, Paul says. "The elites are terrified that peace may finally break out, which will be bad for their profits."

According to a study by the Worldwatch Institute, redirecting just 15 percent of the money spent on weapons globally could eradicate most of the immediate causes of war and environmental destruction.

Despite a growing litany of despoliation, damage, and dangers, Planet Earth might still escape the looming fate of Terracide—but only if we act on Dave Brower's 1982 call to defend "the trees and plants, the isles, the air and water, and . . . the land itself." To do this, we need to dismantle the smoking, oil-fueled engines of Permawar.

(With thanks to Susan D. Lanier-Graham.)

Patriarchy and War: Treating Nature Like Dirt

***Dr. Vandana Shiva, International Forum on Globalization* (2016)**

Wars against the Earth and violence against women are related. They have the same roots in a violent patriarchal worldview where economic creation and production favor the violent conquest of nature and subjugation of women; where everything is transformed into an object of exploitation. Once fundamental holistic connections are broken, separation and fragmentation become the basis of knowledge. In such a world, the violent destruction of nature and nature-based cultures is justified as human "progress."

History, as dictated by male scholars, recounts a succession of wars won and treats violence as something intrinsic to human nature—rather than as a trait intentionally cultivated by patriarchy. Women archaeologists, on the other hand, have shown that through most of human history society has been women-centered, nature-centered, and nonviolent. In countries where women enjoy greater equality and freedom, the general welfare is enriched for everyone—and nature blossoms as well. The focus shifts to sustainability and stability, to compassion and nurturing.

In *The Civilization of the Goddess,* Marija Gimbutas notes how, until around 6,500 years ago, Old Europe cultures were female-centered, peace-loving, communal agricultural societies in which women were respected, even revered. In these ancient cultures, the "goddess" was a divine force that honored the powers of nature, plants, animals, and humans. "With the inception of agriculture," Gimbutas writes, humankind "began to observe the phenomena of the miraculous Earth more closely and more intensively. . . . A separate deity emerged, the Goddess of vegetation, a symbol of the sacral nature of the seed and the sown field, whose ties with the Great Goddess are intimate." In these Old Europe societies, there was no polarizing division between female and male nor any subordination of female to male (a distinction that we subsequently have been taught to accept as "natural").

In contrast to the narrative of "man: the hunter, gatherer, protector," in these ancient societies, it was women who were the innovators and organizers. The main concerns in these Old Europe "goddess" societies were not war and conquest but the protection and regeneration of life. "War was not on any indigenous agenda," Gimbutas explains, "since all human safety and well-being was dependent upon good social relations with one another and with neighbors." Even today, there are forager communities and matriarchal societies in which war is unknown and violence is uncommon.

In the excavations of many ancient tombs and cities, no weapons of war have been found—no swords, no spears, no bows, no arrows. No weapons were found among the artifacts uncovered in the 4,500-year-old ruins of Mohenjo Daro, a "lost city" located in the floodplains of Pakistan. In these Neolithic settlements there were no fortifications to protect villagers from enemies; no signs of violent deaths; no oversized central buildings indicating a hierarchy or dominant ruler. Old Europe was, "in the main, peaceful, sedentary, matrifocal, matrilinear and sexegalitarian."

The idea of being "civilized" necessarily implies that, if one is not "civilized," then one must be "barbaric" or "primitive." Conveniently, all earth-based cultures have been swept into the latter category while all earth-plundering cultures have been proclaimed "civilized." Many peaceful and prosperous human societies that continue to thrive in balance with nature have been labeled "primitive." Meanwhile, the myth of "progress"—implying radical change, transformation, production, and consumption—frees "civilized" nations to commit extraordinary violence in the name of "development," "growth," and "globalization."

As French feminist Simone de Beauvoir has observed: "It is men that form armies and fight wars. It is men who build factories, smelt gold and fell forests. In a patriarchy, men dominate both women and nature. A patriarchal Warrior Culture treats Mother Nature the same way it treats any other female figure."

The word "nature" is derived from the Latin verb *nasci*, "to be born." So it follows that patriarchy objectifies both women and nature and defines the creativity inherent in both as inert and passive.

As Swiss anthropologist Johann Jakob Bachofen writes in *Myth, Religion, and Mother Right*: "The triumph of paternity brings with it the liberation of the spirit from the manifestations of nature. . . . Maternity pertains to the physical side of man, the only thing he shares with animals:

the paternal spiritual principle belongs to him alone. Triumphant paternity partakes of the heavenly light, while child-bearing motherhood is bound up with the earth that bears all things."

This patriarchal construction rests on three separations: the separation of mind and body; the separation of male activity as intellectual, spiritual, and creative; and the separation of female activity as "merely" biological. The separation of production from reproduction—and the characterization of the former as an economic force that creates "value" while the latter is simply biological—has served to institutionalize a false "creation boundary." Today, patriarchal corporations invoke this creation boundary as a rationale to patent life, control seeds, clear-cut forests, strip-mine mountains, drill oil wells, and expropriate traditional knowledge under the claim of "intellectual property rights."

Patriarchal men dominate women by raping them. Patriarchal men dominate nature in the same way. Rip it loose; tear it open; pull it down; bludgeon it; cut it; burn it; make a buck off of it and . . . move on to the next conquest. Modern industrial capitalism has no regard or reverence for nature. Patriarchies view nature as little more than an exhaustible mine of raw materials or a dumping ground for industrial waste. Before the Industrial Revolution, "resources" meant "that which resurges and renews." After the Industrial Revolution, the word was conveniently redefined to mean "raw materials."

Patriarchal science, which arose during the Scientific Revolution of the sixteenth century, helped lay the intellectual foundations of the Industrial Revolution, which further diminished the primacy of nature and women. The violence, the fragmentation, the mechanistic thought, the obsession with uniformity and groupthink, and the will to control that characterize industrial society are all rooted in a patriarchal scientific paradigm. In the words of Sir Francis Bacon (commonly referred to as the "*father* of modern science") the mechanical inventions that flow from scientific research do not "merely exert a gentle guidance over nature's course; they have the power to conquer and subdue her, to shake her to her foundations." Bacon went further in *Temporis Partus Masculus* (*The Masculine Birth of Time*), when he promised to create "a blessed race of heroes and supermen" that would dominate both nature and society.

In 1664, Bacon's philosophy inspired Henry Oldenburg, secretary of Britain's Royal Society, to announce the organization's intention to "raise a masculine philosophy . . . whereby the Mind of the Man may be ennobled with the knowledge of solid truths." And cleric Joseph Glanvill proposed the masculine aim of science was to discover "the ways of captivating Nature, and making her subserve our purposes and designments," thereby achieving the biblical goal of enforcing "the Empire of Man Over Nature."

This was a far cry from the sustainable nature-centric systems we now refer to as the Sustenance Economy, the Subsistence Economy, the Caring Economy, or the Gift Economy. As Carolyn Merchant points out in *The Death of Nature*, this transformation of nature from a living, nurturing mother to inert, dead, and manipulable matter was eminently suited to the exploitation imperative of growing capitalism. After all, if the Earth is merely dead matter, then nothing is being killed. The image of a nurturing Mother Earth acted as a cultural constraint on the exploitation of nature, Merchant notes, since "one does not readily slay a mother, dig her entrails or mutilate her body." But the domination images created by Bacon and the Scientific Revolution removed all restraint, opening the doors for the denudation of nature.

As Ronnie Lessem and Alexander Schieffer write in their book *Integral Economics*: "If the fathers of capitalist theory had chosen a mother rather than a single bourgeois male as the smallest economic unit for their theoretical constructions, they would not have been able to formulate the axiom of the selfish nature of human beings in the way they did." On such false assumptions does the entire edifice of the dominant economic paradigm rest.

Today, who or what counts as a human being is changing. Beginning with the idea of a bourgeois male as the normative "human," patriarchal economics now defines the "corporation" as a patriarchal "person." Seeds, food, and agriculture—all traditional spheres of women's knowledge and production—now are seen primarily as sources of mega-profits for privateering corporations.

Patriarchal science promotes patriarchal economics—profit-before-people systems that feed global corporations and marginalize women. Patriarchal economies exclude women's work because it is undertaken for sustenance, not for profiteering. Women's work in food and agriculture is discounted, even though it is fundamental to human survival. Under the patriarchal productivity calculus, the value of sustainable food systems—shaped by women to support their families, communities, and nature's biodiversity—is reduced to zero.

The transformation of value into "cost," labor into "idleness," knowledge into "ignorance" is achieved by the patriarchal construct of the gross domestic product (which some commentators now call the "gross domestic problem.") The GDP, a national accounting system used to calculate growth, is based on the assumption that if producers consume what they produce, they do not, in fact, produce anything at all. This assumption excludes all regenerative and renewable production cycles from the area of production. Hence, any work that supports families, children, community, and society is treated as "non-productive" and economically "inactive." In short, economic self-sufficiency is perceived as economic deficiency. The devaluation of women's work—and of work done in subsistence economies of the Global South—is the inevitable outcome of a capitalist patriarchy.

The patriarchal market economy ignores two vital forces that are essential to ecological and human survival—nature's economy and the sustenance economy, where economic value is a measure of how human well-being and planetary health are protected. Its currency is life-giving processes, not cash or the market price. Denying women's work and wealth creation only deepens the violence by displacing women from their livelihoods and the natural resources on which their livelihoods depend—land, forests, water, seeds, and nature's biodiversity.

Economic programs based on the fiction of limitless growth in a limited world can only be maintained by the powerful grabbing the resources of the vulnerable. The resource grab that is essential for "growth" creates a culture of rape—of the Earth, of local self-reliant economies, of women. The only way in which such "growth" is "inclusive" is by its inclusion of ever-larger numbers in its circle of violence.

An economics of commodification creates a culture where everything has a price and nothing has value. Economic liberalism has unleashed a flood of commercial deregulation along with the privatization and commodification of seeds, food, land, water, men, women, and children. This has further degraded social values, deepened patriarchy, and intensified violence against women. The growing global scourge of aggressive, male-driven brutality is a social externality of an economics—and a culture—based on dominance, control, competition, and warfare. In short, a definition of "progress" that demands nothing less than the raping of the Earth and its people.

Industrial agriculture is rooted in the same patriarchal scientific paradigm that privileges violence, fragmentation, and mechanistic thought. Rooted in the ideologies and instruments of war, this paradigm promotes "monocultures of the mind" and monocultures on our land that deny the nature-rooted wisdom of agroecology and biodiversity. Herbicides used in industrial agriculture are also used as weapons of war. Chemicals used to kill people in World War II gas chambers were later developed into pesticides. Factories that used to make explosives for the military now make ammonium nitrate fertilizers (which can be used to make bombs, like the one that destroyed the Oklahoma City Federal Building in 1995).

While global corporations have used the foundations laid by masculine science to obscure women's knowledge and productivity, a 1998 Food and Agriculture Organization report has documented how successfully "Women Feed the World." Women are the biodiversity experts of the planet. In Nigeria, women routinely harvest eighteen to fifty-seven plant species in a single home garden. In sub-Saharan Africa, women cultivate as many as 120 different plants in the spaces left alongside the cash crops managed by men. In Guatemala, home gardeners grow more than ten tree and crop species on one-tenth of a hectare. A single African home garden can host more than sixty species of food-producing trees. In India, women are using 150 different species of plants for vegetables, fodder, and health care.

While women manage and protect biodiversity, the dominant paradigm of industrial agriculture promotes monoculture on the false assumption that single-crop farming produces more food. But monocultures do *not* produce more—they simply concentrate control and power in the hands of a few. Since women's expertise is based on modeling agriculture on nature's methods of renewability, the suppression of this knowledge has gone hand in hand with the ecological destruction of nature's processes—and the destruction of people's livelihoods and lives.

Since the mid-fourteenth century, the word "rape" has meant "the act of taking something by force, plundering." Rape, in its original meaning, remains an accurate description of today's globalized economy—a system based on plunder of resources through biopiracy, land grabs, water theft, seed patents, the creation of genetically engineered life-forms, and an intensified global war against the living Earth.

Could there be a connection between the growth of violent, undemocratically imposed, unjust, and unfair economic policies and the intensification of brutal crimes against women? I believe there is. As the rape of the Earth intensifies, so does violence against women. I am not suggesting that violence against women begins with neoliberal economics. I am deeply aware of the profound gender biases in our traditional cultures and social organizations. (I stand empowered today because my grandfather sacrificed his life for women's equality and my mother was a feminist before the word existed.) However, it is clear that violence against women has taken on new and more vicious and brutal forms as traditional patriarchal structures have hybridized with the structures of capitalist patriarchy. We need to examine the connections between unjust, non-sustainable economic systems and the growing frequency and brutality of violence against women.

The ruling paradigm that falsely defines our economy as a "market" imposed on us in the name of "growth" fuels the intensity of crimes against women while ravaging the Earth and deepening social and economic inequality. Social and economic reforms can no longer be insulated from each other. We need to move beyond the violent economy shaped by capitalist patriarchy to new nonviolent, sustainable, and peaceful economies that give respect to women and the Earth.

U.S. Exceptionalism: The Hubris That Fuels Wars

Marjorie Cohn (2015)

On April 23, 2015, Barack Obama stood behind the presidential podium and apologized for inadvertently killing two Western hostages during a drone strike in Pakistan. "One of the things that sets America apart from many other nations—one of the things that makes us exceptional—is our willingness to confront squarely our imperfections and to learn from our mistakes," he said.

In his 2015 State of the Union address, Obama again described America as "exceptional." When he spoke to the United Nations General Assembly in 2013, he said, "Some may disagree, but I believe that America is exceptional."

American exceptionalism reflects the belief that Americans are somehow better than everyone else. This view reared its head after the 2013 leak of a Justice Department white paper that describes circumstances under which the president can order the targeted killing of U.S. citizens. There had been little public concern in this country about drone strikes that killed people in other countries. But when it was revealed that U.S. citizens could be targeted, Americans were outraged. This motivated Senator Rand Paul (R-Kentucky) to launch a thirteen-hour filibuster of John Brennan's nomination to become the next CIA director.

It is this double standard that moved Nobel Peace Prize–winner Archbishop Desmond Tutu to write a letter to the editor of the *New York Times*, in which he asked, "Do the United States and its people really want to tell those of us who live in the rest of the world that our lives are not of the same value as yours?"

Obama insists that the CIA and the Pentagon are careful to avoid civilian casualties. In May 2013, he declared in a speech at the National Defense University: "Before any strike is taken, there must be near-certainty that no civilians will be killed or injured—the highest standard we can set."

Nevertheless, according to the Bureau of Investigative Journalism's ongoing tally of drone casualties (as of August 2015), of the 5,403 people killed by drone strikes in Afghanistan, Somalia, Pakistan, and Yemen, 1,109 were reported to have been civilians and 234 were children. The Open Society Justice Initiative (OSJI), which examined nine drone strikes in Yemen, concluded that civilians were killed in every one. Amrit Singh, a senior legal officer at OSJI and primary author of the report, said: "We've found evidence that President Obama's standard is not being met on the ground."

In 2013, the administration released a fact sheet specifying that, in order to use lethal force, the target must pose a "continuing, imminent threat to U.S. persons." But the leaked Justice Department white paper says that a U.S. citizen can be killed even when there is no "clear evidence that a specific attack on U.S. persons and interests will take place in the immediate future."

There must also be "near certainty" that the terrorist target is present. Yet the CIA did not even know who it was slaying when it killed two hostages (an Italian and an American) in an attack on al-Qaeda in Pakistan in January 2015. This was a "signature strike" that targeted "suspicious compounds" in areas controlled by "militants."

Most individuals killed are not on a kill list and the president does not know their names. So how can one determine with any certainty that a target is present when the CIA is not even targeting individuals?

Contrary to popular opinion, the use of drones does not result in fewer civilian casualties than manned bombers. A study based on classified military data, conducted by the Center for Naval Analyses and the Center for Civilians in Conflict, concluded that the use of drones in Afghanistan caused ten times more civilian deaths than manned fighter aircraft.

Moreover, a panel with experienced specialists from both the George W. Bush and Bill Clinton administrations issued a 77-page report for the Stimson Center, a nonpartisan think tank, which found there was no indication that drone strikes had advanced "long-term U.S. security interests."

Nevertheless, the Obama administration maintains a double standard for apologies to the families of drone victims. "The White House is setting a dangerous precedent—that if you are Western and hit by accident, we'll say we are sorry," said Reprieve attorney Alka Pradhan, "but we'll put up a stone wall of silence if you are a Yemeni or Pakistani civilian who lost an innocent loved one. Inconsistencies like this are seen around the world as hypocritical, and do the United States' image real harm."

It is not just the U.S. image that is suffering. Drone strikes create more enemies of the United States. While Faisal Shahzad was pleading guilty to trying to detonate a bomb in Times Square, he told the judge, "When the drones hit, they don't see children."

In 2009, former CIA lawyer Vicki Divoll, who now teaches at the U.S. Naval Academy, told the *New Yorker*'s Jane Mayer: "People are a lot more comfortable with a Predator [drone] strike that kills many people than with a throat-slitting that kills one." But Americans don't see the images of the drone victims or hear the stories of their survivors. If we did, we might be more sympathetic to the damage our drone bombs are wreaking in our name.

The guarantee of due process in the U.S. Constitution as well as in the International Covenant on Civil and Political Rights must be honored. That means arrest and fair trial, not summary execution.

What we really need is a complete reassessment of the "war on terror." Until we overhaul our foreign policy and stop invading other countries, changing their regimes, occupying, torturing, and indefinitely detaining their people—and uncritically supporting other countries that illegally occupy other peoples' lands—we will never be safe from terrorism.

Blowback: Climate Change and Resource Wars

Michael T. Klare (2013)

Brace yourself. You may not be able to tell yet, but according to global experts and the U.S. intelligence community, the Earth is already shifting under you. Whether you know it or not, you're on a new planet, a resource-shock world of a sort humanity has never before experienced.

Two nightmare scenarios—a global scarcity of vital resources and the onset of extreme climate change—are already beginning to converge and, in the coming decades, are likely to produce a tidal wave of unrest, rebellion, competition, and conflict. Just what this tsunami of disaster will look like may, as yet, be hard to discern, but experts warn of "water wars" over contested river systems, global food riots sparked by soaring prices for life's basics, mass migrations of climate refugees (with resulting anti-migrant violence), and the breakdown of social order or the collapse of states. At first, such mayhem is likely to arise largely in Africa, Central Asia, and other areas of the underdeveloped South, but in time, *all* regions of the planet will be affected.

Start with one simple given: the prospect of future scarcities of vital natural resources, including energy, water, land, food, and critical minerals. This in itself would guarantee social unrest, geopolitical friction, and war.

It is important to note that absolute scarcity doesn't have to be on the horizon in any given resource category for this scenario to kick in. A lack of adequate supplies to meet the needs of a growing, ever more urbanized and industrialized global population is enough. Given the wave of extinctions that scientists are recording, some resources—particular species of fish, animals, and trees, for example—will become less abundant in the decades to come and may even disappear altogether. But key materials for modern civilization like oil, uranium, and copper will simply prove harder and more costly to acquire.

Oil—the single most important commodity in the international economy—provides an apt example. In its 2011 *World Energy Outlook*, the International Energy Agency claimed that an anticipated global oil demand of 104 million barrels per day in 2035 will be satisfied, thanks, in large part, to additional supplies of "unconventional oil" (Canadian tar sands, shale oil, and so on), as well as 55 million barrels of new oil from fields "yet to be found."

However, many analysts scoff at this optimistic assessment, arguing that rising production costs, environmental opposition, warfare, corruption, and other impediments will make it extremely difficult to achieve increases of this magnitude.

Water provides another potent example. On an annual basis, the supply of drinking water provided by natural precipitation remains more or less constant: about 40,000 cubic kilometers [9,597 cubic miles]. But much of this precipitation lands on Greenland, Antarctica, Siberia, and inner Amazonia, so the supply available to major concentrations of humanity is often surprisingly limited. In many regions with high population levels, water supplies are already relatively sparse. This is especially true of North Africa, Central Asia, and the Middle East, where the demand for water continues to grow as a result of rising populations, urbanization, and the emergence of new water-intensive industries.

Wherever you look, the picture is roughly the same: supplies of critical resources may be rising or falling, but rarely do they appear to be outpacing demand, producing a sense of widespread and systemic scarcity. However generated, a perception of scarcity—or imminent scarcity—regularly leads to anxiety, resentment, hostility, and contentiousness. This pattern has been evident throughout human history. In his book *Constant*

Battles, Steven LeBlanc, director of collections for Harvard's Peabody Museum of Archaeology and Ethnology, notes that many ancient civilizations experienced higher levels of warfare when faced with resource shortages brought about by population growth, crop failures, or drought. Jared Diamond, author of *Collapse: How Societies Choose to Fail or Succeed*, has detected a similar pattern in Mayan civilization and the Anasazi culture of New Mexico's Chaco Canyon. According to Lizzie Collingham, author of *The Taste of War*, concern over adequate food was a significant factor in Japan's invasion of Manchuria in 1931 and Germany's invasions of Poland in 1939 and the Soviet Union in 1941.

Although the global supply of most basic commodities has grown enormously since the end of World War II, analysts see the persistence of resource-related conflict in areas where materials remain scarce or there is anxiety about the future reliability of supplies. Many experts believe, for example, that the fighting in Darfur and other war-ravaged areas of North Africa has been driven, at least in part, by competition among desert tribes for access to scarce water supplies, exacerbated in some cases by rising population levels.

"In Darfur," says a 2009 report from the UN Environment Programme, "recurrent drought, increasing demographic pressures, and political marginalization are among the forces that have pushed the region into a spiral of lawlessness and violence that has led to 300,000 deaths and the displacement of more than two million people since 2003."

Anxiety over future supplies is often also a factor in conflicts that break out over access to oil or control of contested undersea reserves of oil and natural gas. In 1979, for instance, when the Islamic Revolution in Iran overthrew the shah and the Soviets invaded Afghanistan, Washington began to fear that someday it might be denied access to Persian Gulf oil. In his 1980 State of the Union Address, President Jimmy Carter affirmed that any move to impede the flow of oil from the Gulf would be viewed as a threat to America's "vital interests" and would be repelled by "any means necessary, including military force."

In 1990, this "Carter Doctrine" was invoked by President George H. W. Bush to justify intervention in the first Persian Gulf War, just as his son would use it, in part, to justify the 2003 invasion of Iraq. It remains the basis for U.S. plans to employ force to stop Iran from closing the Strait of

Hormuz, the strategic waterway connecting the Persian Gulf to the Indian Ocean through which about 35 percent of the world's seaborne oil commerce passes.

Recently, a set of resource conflicts have been rising toward the boiling point between China and its neighbors in Southeast Asia when it comes to control of offshore oil and gas reserves in the South China Sea. A similar situation has also arisen in the East China Sea, where China and Japan are jousting for control over similarly valuable undersea reserves. Meanwhile, in the South Atlantic Ocean, Argentina and Britain are once again squabbling over the Falkland Islands (called Las Malvinas by Argentina) because oil has been discovered in surrounding waters.

By all accounts, resource-driven potential conflicts like these will only multiply in the years ahead as demand rises, supplies dwindle, and more of what remains will be found in disputed areas.

Heading for a Resource-Shock World

On this planet, a second major force has entered the equation in a significant way. With the growing reality of climate change, everything becomes a lot more terrifying.

Normally, when we consider the impact of climate change, we think primarily about the melting Arctic ice cap or Greenland ice shield, rising global sea levels, intensifying storms, expanding deserts, and endangered or disappearing species like the polar bear. But a growing number of experts are coming to realize that the most potent effects of climate change will be experienced by humans directly through the impairment or wholesale destruction of habitats upon which we rely for the basics of life—food, water, land, and energy.

We already know enough about the future effects of climate change to predict the following with reasonable confidence:

- Rising sea levels will, in the next half-century, erase many coastal areas, destroying large cities, critical infrastructure (including roads, railroads, ports, airports, pipelines, refineries, and power plants) and prime agricultural land.

- Diminished rainfall and prolonged droughts will turn once-verdant croplands into dust bowls, reducing food output and turning millions into "climate refugees."
- More severe storms and intense heat waves will kill crops, trigger forest fires, cause floods, and destroy critical infrastructure.

No one can predict how much food, land, water, and energy will be lost, but the cumulative effect will undoubtedly be staggering. In *Resources Futures*, Chatham House offers a particularly dire warning when it comes to the threat of diminished precipitation to rain-fed agriculture. "By 2020," the report says, "yields from rain-fed agriculture could be reduced by up to 50 percent" in some areas. The highest rates of loss are expected in Africa, where reliance on rain-fed farming is greatest, but agriculture in China, India, Pakistan, and Central Asia is also likely to be severely affected.

Climate change will also reduce the flow of many vital rivers, diminishing water supplies for irrigation, hydroelectricity power, and nuclear reactors (which need massive amounts of water for cooling). The melting of glaciers, especially in the Andes in Latin America and the Himalayas in South Asia, will rob communities and cities of crucial water supplies. An expected increase in the frequency of hurricanes and typhoons will pose a growing threat to offshore oil rigs, coastal refineries, transmission lines, and other components of the global energy system. The melting of polar ice caps will open the Arctic to oil and gas exploration, but an increase in iceberg activity will make efforts to exploit the region's resources perilous and exceedingly costly.

Climate Change: A "Threat Multiplier"

Longer growing seasons in the north, especially in Siberia and Canada's northern provinces, might compensate to some degree for the desiccation of croplands in southerly latitudes. However, moving the global agricultural system (and the world's farmers) northward from abandoned farmlands in the United States, Mexico, Brazil, India, China, Argentina, and Australia would be a daunting prospect.

It is safe to assume that climate change, especially when combined with growing supply shortages, will result in a significant reduction in the planet's vital resources, augmenting the kinds of pressures that have historically led to conflict, even under better circumstances. According to the Chatham House report, climate change is best understood as a "threat multiplier . . . a key factor exacerbating existing resource vulnerability" in states already prone to such disorders. "Increased frequency and severity of extreme weather events, such as droughts, heat waves and floods, will also result in much larger and frequent local harvest shocks around the world. . . . These shocks will affect global food prices whenever key centers of agricultural production area are hit—further amplifying global food price volatility." This, in turn, will increase the likelihood of civil unrest.

When, for instance, a brutal heat wave decimated Russia's wheat crop during the summer of 2010, the global price of wheat (and of that staple of life, bread) began an inexorable upward climb, reaching particularly high levels in North Africa and the Middle East. Anger over impossible-to-afford food merged with resentment toward autocratic regimes to trigger the massive popular outburst we know as the Arab Spring.

In March 2013, for the first time, Director of National Intelligence James R. Clapper listed "competition and scarcity involving natural resources" as a national security threat on a par with global terrorism, cyberwar, and nuclear proliferation.

There was a new phrase embedded in his comments: "resource shocks." It catches something of the world we're barreling toward, and the language is striking for an intelligence community that, like the government it serves, has largely played down or ignored the dangers of climate change. For the first time, senior government analysts may be coming to appreciate what energy experts, resource analysts, and scientists have long been warning about: the unbridled consumption of the world's natural resources, combined with the advent of extreme climate change, could produce a global explosion of human chaos and conflict. We are now heading directly into a resource-shock world.

War Is Not Biological

Margaret Mead, American Anthropological Association (1940)

Is war a biological necessity, a sociological inevitability, or just a bad invention? Those who argue for the first view endow man with such pugnacious instincts that some outlet in aggressive behavior is necessary if man is to reach full human stature. It was this point of view, which lay behind William James's famous essay, "The Moral Equivalent of War," in which he tried to retain the warlike virtues and channel them in new directions.

A basic, competitive, aggressive, warring human nature is assumed, and those who wish to outlaw war or outlaw competitiveness merely try to find new and less socially destructive ways in which these biologically given aspects of man's nature can find expression.

Then there are those who take the second view: warfare is the inevitable concomitant of the development of the state, the struggle for land and natural resources, of class societies springing not from the nature of man, but from the nature of history. War is nevertheless inevitable unless we change our social system and outlaw classes, the struggle for power, and possessions; and in the event of our success, warfare would disappear, as a symptom vanishes when the disease is cured.

One may hold a sort of compromise position between these two extremes; one may claim that all aggression springs from the frustration of man's biologically determined drives and that, since all forms of culture are frustrating, it is certain each new generation will be aggressive and the aggression will find its natural and inevitable expression in race war, class war, nationalistic war, and so on. All three of these positions are very popular today among those who think seriously about the problems of war and its possible prevention, but I wish to urge another point of view, less defeatist, perhaps, than the first and third and more accurate than the second: that is, that warfare—by which I mean recognized conflict between two groups as groups, in which each group puts an army (even if the army is only fifteen pygmies) into the field to fight and kill, if possible, some of the members of the army of the other group—is an invention like any other of the inventions in terms of which we order our lives, such as writing, marriage, cooking our food instead of eating it raw, trial by jury, or burial of the dead, and so on.

Whenever a way of doing things is found universally, such as the use of fire or the practice of some form of marriage, we tend to think at once that it is not an invention at all but an attribute of humanity itself. And yet, even such universals as marriage and the use of fire are inventions like the rest, very basic ones, inventions which were, perhaps, necessary if human history was to take the turn that it has taken, but nevertheless inventions. At some point in his social development, man was undoubtedly without the institution of marriage or the knowledge of the use of fire.

The case for warfare is much clearer because there are peoples even today who have no warfare. Of these, the Eskimos are perhaps the most conspicuous examples, but the Lepchas of Sikkim are as good. Neither of these peoples understands war, not even defensive warfare. The idea of warfare is lacking, and this idea is as essential to really carrying on war as an alphabet or a syllabary is to writing.

But, whereas the Lepchas are a gentle, unquarrelsome people, the Eskimo case gives no such possibility of interpretation. The Eskimos are not a mild and meek people; many of them are turbulent and troublesome. Fights, theft of wives, murder, cannibalism occur among them—all outbursts of passionate men goaded by desire or intolerable circumstance. Here are men faced with hunger, men faced with loss of their wives, men faced with the threat of extermination by other men, and here are orphan children, growing up miserably with no one to care for them, mocked and neglected by those about them.

The personality necessary for war, the circumstances necessary to goad men to desperation are present, but there is no war. When a traveling Eskimo entered a settlement, he might have to fight the strongest man in the settlement to establish his position among them, but this was a test of strength and bravery, not war. The idea of warfare, of one group organizing against another group to maim and wound and kill them, was absent. And, without that idea, passions might rage, but there was no war.

But, it may be argued, is not this because the Eskimos have such a low and undeveloped form of social organization? They own no land, they move from place to place, camping, it is true, season after season on the same site, but this is not something to fight for as the modern nations of the world fight for land and raw materials. They have no permanent possessions that can be looted, no towns that can be burned. They have no social classes to produce stress and strains within the society, which might force it to go to war outside. Does not the absence of war among the Eskimos, while disproving the biological necessity of war, just go to confirm the point that it is the state of development of the society that accounts for war and nothing else?

We find the answer among the pygmy peoples of the Andaman Islands in the Bay of Bengal. The Andamans also represent an exceedingly low level of society; they are a hunting and food-gathering people; they live in tiny hordes without any class stratification; their houses are simpler than the snow houses of the Eskimo. But they knew about warfare. The army might contain only fifteen determined pygmies marching in a straight line, but it was the real thing nonetheless. Tiny army met tiny army in open battle, blows were exchanged, casualties suffered, and the state of warfare could only be concluded by a peacemaking ceremony.

Similarly, among the Australian aborigines, who built no permanent dwellings but wandered from water hole to water hole over their almost desert country, warfare—and rules of "international law"—were highly developed. The student of social evolution will seek in vain for his obvious causes of war, struggle for lands, struggle for power of one group over another, expansion of population, need to divert the minds of a populace restive under tyranny, or even the ambition of a successful leader to enhance his own prestige. All are absent, but warfare as a practice remained, and men engaged in it and killed one another in the course of a war because killing is what is done in wars.

From instances like these, it becomes apparent that an inquiry into the causes of war misses the fundamental point as completely as does an insistence upon the biological necessity of war. If a people have an idea of going to war and the idea that war is the way in which certain situations, defined within their society, are to be handled, they will sometimes go to war. If they are a mild and unaggressive people, like the Pueblo Indians, they may limit themselves to defensive warfare, but they will be forced to think in terms of war because there are peoples near them who have warfare as a pattern, and offensive, raiding, pillaging warfare at that.

So simple peoples and civilized peoples, mild peoples and violent, assertive peoples will all go to war if they have the invention, just as those peoples who have the custom of dueling will have duels and peoples who have the pattern of vendetta will indulge in vendetta. And, conversely, peoples who do not know of dueling will not fight duels, even though their wives are seduced and their daughters ravished; they may, on occasion, commit murder but they will not fight duels. Cultures which lack the idea of the vendetta will not meet every quarrel in this way.

A people can use only the forms it has. So the Balinese have their special way of dealing with a quarrel between two individuals: if the two feel that the causes of quarrel are heavy, they may go and register their quarrel in the temple before the gods, and, making offerings, they may swear never to have anything to do with each other again.

Yet, if it be granted that warfare is, after all, an invention, it may nevertheless be an invention that lends itself to certain types of personality, to the exigent needs of autocrats, to the expansionist desires of crowded peoples, to the desire for plunder and rape and loot which is engendered by a dull and frustrating life.

There are tribes who go to war merely for glory, having no quarrel with the enemy, suffering from no tyrant within their boundaries, anxious neither for land nor loot nor women, but merely anxious to win prestige which within that tribe has been declared obtainable only by war and without which no young man can hope to win his sweetheart's smile of approval. But if, as was the case with the Bush Negroes of Dutch Guiana, it is artistic ability which is necessary to win a girl's approval, the same young man would have to be carving rather than going out on a war party.

In many parts of the world, war is a game in which the individual can win counters—counters which bring him prestige in the eyes of his own sex or of the opposite sex; he plays for these counters as he might, in our society, strive for a tennis championship.

Warfare is a frame for such prestige-seeking merely because it calls for the display of certain skills and certain virtues; all of these skills—riding straight, shooting straight, dodging the missiles of the enemy and sending one's own straight to the mark—can be equally well exercised in some other framework and, equally, the virtues endurance, bravery, loyalty, steadfastness can be displayed in other contexts.

The tie-up between proving oneself a man and proving this by a success in organized killing is due to a definition which many societies have made of manliness. Warfare is just an invention known to the majority of human societies by which they permit their young men either to accumulate prestige or avenge their honor or acquire loot or wives or slaves or sago lands or cattle or appease the bloodlust of their gods or the restless souls of the recently dead. It is just an invention, older and more widespread than the jury system, but nonetheless an invention.

But, once we have said this, have we said anything at all? Grant that war is an invention—that it is not a biological necessity nor the outcome of certain special types of social forms—still, once the invention is made, what are we to do about it? Once an invention is known and accepted, men do not easily relinquish it. The skilled workers may smash the first steam looms which they feel are to be their undoing, but they accept them in the end, and no movement which has insisted upon the mere abandonment of usable inventions has ever had much success.

Warfare is here, as part of our thought: the deeds of warriors are immortalized in the words of our poets; the toys of our children are modeled upon the weapons of the soldier; the frame of reference within which our statesmen and our diplomats work always contains war. If we know that it is not inevitable, that it is due to historical accident that warfare is one of the ways in which we think of behaving, are we given any hope by that? What hope is there of persuading nations to abandon war, nations so thoroughly imbued with the idea that resort to war is, if not actually desirable and noble, at least inevitable whenever certain defined circumstances arise?

In answer to this question, I think we might turn to the history of other social inventions and inventions, which must once have seemed as finally entrenched as warfare. Take the methods of trial, which preceded the jury system: ordeal and trial by combat. Unfair, capricious, alien as they are to our feeling today, they were once the only methods open to individuals accused of some offense. The invention of trial by jury gradually replaced these methods until only witches, and finally not even witches, had to resort to the ordeal. The ordeal did not go out because people thought it unjust or wrong; it went out because a method more congruent with the institutions and feelings of the period was invented. And, if we despair over the way in which war seems such an ingrained habit of most of the human race, we can take comfort from the fact that a poor invention will usually give place to a better invention.

For this, two conditions, at least, are necessary. The people must recognize the defects of the old invention, and someone must make a new one. Propaganda against warfare, documentation of its terrible cost in human suffering and social waste, these prepare the ground by teaching people to feel that warfare is a defective social institution. There is further needed a belief that social invention is possible and the invention of new methods which will render warfare as out-of-date as the tractor is making the plow, or the motorcar the horse and buggy. A form of behavior becomes out of date only when something else takes its place, and, in order to invent forms of behavior, which will make war obsolete, it is a first requirement to believe that an invention is possible.

Lessons from the Bonobos

Sally Jewell Coxe (2016)

Deep in the heart of the Congo, legends linger about an elusive, "almost human" shadow—a mysterious ape, shrouded from the allure accorded its cousins, the chimpanzees, gorillas, and orangutans. The bonobo (*Pan paniscus*) is the great ape most closely related to humans, sharing 98.4 percent of our DNA and displaying many qualities that we humans need to emulate to ensure our own survival, and that of our planet.

Bonobos were the last great ape to be studied by modern scientists and, unless protections are enforced immediately, they could be the first to go extinct. These rare apes inhabit the central Congo Basin—home to the world's second-largest rainforest and the area of greatest biodiversity in Africa. Found only in the Democratic Republic of the Congo (DRC), a resource-rich region ravaged by civil war and foreign occupation, bonobos face an ironic fate. Distinguished by their peaceful, loving nature, bonobos have become victims of human violence.

Unlike their close relatives, the chimpanzees (*Pan troglodytes*)—who engage in a competitive, male-dominated society and wage territorial wars against each other—bonobos have a matriarchal culture, bound by cooperation, sharing, and the creative use of sex. Bonobos live in large groups where peaceful coexistence is the norm. Females carry the highest rank and the sons of ranking females are the leaders among males. Alliances among females are the central unifying force.

Bonobos show how a complex society can be ordered successfully by cooperation rather than competition. When neighboring groups of bonobos meet in the forest, they greet one another sexually and share food instead of fighting. Dubbed the "hippie chimps," bonobos exemplify the 1960s credo, "Make love, not war." Sex transcends reproduction in bonobos (as it does in humans). Bonobos are bisexual or, as psychologist Frans de Waal contends, "pansexual." Sex permeates almost all aspects of daily life. Encounters, both with the same and the opposite sex, serve as a way of bonding, sharing, and keeping the peace. Unlike other apes, bonobos frequently copulate face to face, looking into each other's eyes.

Bonobo anatomy is strikingly similar to that of our early human ancestor, *Australopithecus*. Bonobos walk bipedally more easily and more often than other apes. The uncommon social structure, sexual behavior, and intellectual capacity of bonobos reveal compelling clues about the roots of human nature. Highly compassionate and conscious beings, bonobos blur the line between animal and human.

Much of what we know about the bonobo mind and emotion we owe to Kanzi and his sister Panbanisha, who currently live at the Georgia State University Language Research Center near Atlanta. Under the tutelage of Dr. Sue Savage-Rumbaugh, these bonobos have learned to understand spoken English and can communicate using sign language. The bonobos "speak" by pointing to lexigrams or symbols on a keyboard that correspond to words.

Kanzi and Panbanisha have been my best teachers, and they have inspired my work for bonobo conservation. The first day I had direct contact with Panbanisha several years ago, we went for a walk in the forest surrounding the lab. When we stopped to rest and have a snack, Panbanisha began to groom me, combing my hair with her fingers, inspecting the contours of my face.

When she discovered a cut on my wrist, she pointed to it, furrowed her brow and made soft "whu" sounds with a doctorly air of concern. Then she said "hurt" on her keyboard. Once she was convinced that this "hurt" was not "bad," she proceeded to bite my fingernails! Quite the manicurist, Panbanisha peeled a twig, making a sharp point that she used to clean under what remained of my nails, carefully attending to each finger, one by one. This is bonding, bonobo-style. I was honored to be accepted by Panbanisha and as happy as she was to have made a new friend.

In the wild, it is clear that bonobos have a complex communication system, which they use to coordinate their movements through the forest, breaking into small groups for foraging during the day and regrouping at night. When bonobos gather in the trees to make their night nests, they fill the twilight with a symphony of high-pitched soprano squeals that sound like the cries of exotic birds—and quite unlike the guttural hoots of chimpanzees. The Indigenous Mongandu people who live among bonobos still use a "whistle language" to communicate that is eerily reminiscent of bonobo calls.

Unfortunately, owing to the outbreak of regional warfare, bonobos increasingly are being hunted throughout their habitat and little is being done to protect them. The population, small to begin with, is fragmented and decreasing. No one knows how many bonobos survive. Estimates range between 5,000 and 20,000. We do know that bonobos have disappeared from several areas where they formerly lived.

Traditional taboos, which once protected bonobos, are breaking down in the face of economic desperation and human population pressure. More and more bonobos are being killed, both for sustenance and for profit in the commercial bushmeat trade, which is ravaging central Africa.

Our survival as a species may pivot on whether we behave more like chimpanzees or bonobos. Thankfully, there is hope. After years of civil war, the peace process is finally moving forward in the DRC. It is now possible to resume conservation work. As the Congo War abates, concerted efforts can begin to protect bonobos and their habitat and to recognize the apes as a national treasure and icons of peace.

THE BUSINESS OF WAR

Washington cannot tell the American people that the real purpose of its gargantuan military expenditures and belligerent interventions is to make the world safe for General Motors, General Electric, General Dynamics, and all the other generals.

—Michael Parenti, political scientist and author

War Is a Racket

Major-General Smedley Butler (1935)

War is a racket. It always has been.

It is possibly the oldest, easily the most profitable, surely the most vicious. It is the only one international in scope. It is the only one in which the profits are reckoned in dollars and the losses in lives.

There are only two things we should fight for. One is the defense of our homes and the other is the Bill of Rights. War for any other reason is simply a racket.

There isn't a trick in the racketeering bag that the military gang is blind to. It has its "finger men" to point out enemies, its "muscle men" to destroy enemies, its "brain men" to plan war preparations, and a "Big Boss"—Super-Nationalistic-Capitalism.

It may seem odd for me, a military man, to adopt such a comparison. Truthfulness compels me to.

I spent thirty-three years and four months in active military service as a member of this country's most agile military force, the Marine Corps. I served in all commissioned ranks from Second Lieutenant to Major-General. And during that period, I spent most of my time being a high-class muscle man for Big Business, for Wall Street, and for the Bankers. In short, I was a racketeer—a gangster for capitalism.

I suspected I was just part of a racket at the time. Now I am sure of it. Like all the members of the military profession, I never had a thought of my own until I left the service. My mental faculties remained in suspended animation while I obeyed the orders of higher-ups. This is typical with everyone in the military service.

I helped make Mexico, especially Tampico, safe for American oil interests in 1914. I helped make Haiti and Cuba a decent place for the National City Bank boys to collect revenues in. I helped in the raping of half a dozen Central American republics for the benefits of Wall Street. The record of racketeering is long. I helped purify Nicaragua for the international banking house of Brown Brothers in 1909–12. I brought light to the Dominican Republic for American sugar interests in 1916. In China, I helped to see to it that Standard Oil went its way unmolested.

During those years, I had, as the boys in the back room would say, a swell racket. Looking back on it, I feel that I could have given Al Capone a few hints. The best he could do was to operate his racket in three districts. I operated on three continents.

A racket is best described, I believe, as something that is not what it seems to the majority of the people. Only a small "inside" group knows what it is about. It is conducted for the benefit of the very few, at the expense of the very many. Out of war a few people make huge fortunes.

In the World War, a mere handful garnered the profits of the conflict. At least 21,000 new millionaires and billionaires were made in the United States during the World War. That many admitted their huge blood gains in their income tax returns. How many other war millionaires falsified their tax returns no one knows.

How many of these war millionaires shouldered a rifle? How many of them dug a trench? How many of them knew what it meant to go hungry in a rat-infested dug-out? How many of them spent sleepless, frightened nights, ducking shells and shrapnel and machine-gun bullets? How many of them parried a bayonet thrust of an enemy? How many of them were wounded or killed in battle?

Out of war, nations acquire additional territory, if they are victorious. They just take it. This newly acquired territory promptly is exploited by the few—the selfsame few who wrung dollars out of blood in the war. The general public shoulders the bill.

And what is this bill? This bill renders a horrible accounting. Newly placed gravestones. Mangled bodies. Shattered minds. Broken hearts and homes. Economic instability. Depression and all its attendant miseries. Backbreaking taxation for generations and generations.

Now that I see the international war clouds gathering, as they are today, I must face it and speak out. Millions and billions of dollars would be piled up. By a few. Munitions makers. Bankers. Shipbuilders. Manufacturers. Meat packers. Speculators. They would fare well.

Yes, they are getting ready for another war. Why shouldn't they? It pays high dividends.

But what does it profit the men who are killed? What does it profit their mothers and sisters, their wives and their sweethearts? What does it profit their children? What does it profit anyone except the very few to whom war means huge profits?

Yes, and what does it profit the nation? It has been estimated by statisticians and economists and researchers that the war cost your Uncle Sam $52 trillion. Of this sum, $39 trillion was expended in the actual war itself. This expenditure yielded $16 trillion in profits. That is how the 21,000 billionaires and millionaires got that way. And it went to a very few.

Who Pays the Bills?

The bankers collected their profits. But the soldier pays the biggest part of the bill.

Boys with a normal viewpoint were taken out of the fields and offices and factories and classrooms and put into the ranks. There they were remolded; they were made over; they were made to "about face"; to regard murder as the order of the day. They were put shoulder to shoulder and, through mass psychology, they were entirely changed. We used them for a couple of years and trained them to think nothing at all of killing or of being killed.

Then, suddenly, we discharged them and told them to make another "about face!" This time, they had to do their own readjustment, sans mass psychology, sans officers' aid and advice and sans nationwide propaganda. We didn't need them anymore. So we scattered them about without any "three-minute" or "Liberty Loan" speeches or parades. Many, too many, of these fine young boys are eventually destroyed, mentally, because they could not make that final "about face" alone. . . .

In the World War, we used propaganda to make the boys accept conscription. They were made to feel ashamed if they didn't join the army. So vicious was this war propaganda that even God was brought into it. With few exceptions, our clergymen joined in the clamor to kill, kill, kill. To kill the Germans. God is on our side . . . it is His will that the Germans be killed.

And in Germany, the good pastors called upon the Germans to kill the allies . . . to please the same God. That was a part of the general propaganda, built up to make people war-conscious and murder-conscious.

To Hell with War!

Woodrow Wilson was re-elected president in 1916 on a platform that he had "kept us out of war" and on the implied promise that he would "keep us out of war." Yet, five months later, he asked Congress to declare war on Germany. In that five-month interval, the people had not been asked whether they had changed their minds. The four million young men who put on uniforms and marched or sailed away were not asked whether they wanted to go forth to suffer and die.

Then what caused our government to change its mind so suddenly?

Money.

An allied commission, it may be recalled, came over shortly before the war declaration and called on the president. The president summoned a group of advisors. The head of the commission spoke. Stripped of its diplomatic language, this is what he told the president and his group:

"There is no use kidding ourselves any longer. The cause of the allies is lost. We now owe you (American bankers, American munitions makers, American manufacturers, American speculators, American exporters) five or six billion dollars.

"If we lose (and without the help of the United States we must lose), we—England, France and Italy—cannot pay back this money . . . and Germany won't. So . . ."

Had the press been invited to be present at that conference, or had radio been available to broadcast the proceedings, America never would have entered the World War. But this conference, like all war discussions, was shrouded in utmost secrecy. When our boys were sent off to war, they were told it was a "war to make the world safe for democracy" and a "war to end all wars."

Well, eighteen years after, the world has less of democracy than it had then. Besides, what business is it of ours whether Russia or Germany or England or France or Italy or Austria live under democracies or monarchies? Whether they are Fascists or Communists? Our problem is to preserve our own democracy. . . .

The professional soldiers and sailors don't want to disarm. No admiral wants to be without a ship. No general wants to be without a command. Both mean men without jobs. They are not for disarmament. They cannot be for limitations of arms. . . .

There is only one way to disarm with any semblance of practicability. That is for all nations to get together and scrap every ship, every gun, every rifle, every tank, every warplane.

War as an Economic Strategy

Jerry Mander, International Forum on Globalization (2010)

> *War is the health of the state.*
>
> —Randolph Bourne, American journalist and pacifist (1918)

Wars were not invented by capitalism. Historically, plenty of other motives have triggered war—from theology to ideology to avenging insults, recovering land previously taken, resource capture, monarchical or imperial ego and madness, romantic love (Helen of Troy!), and the search for slaves, among others. But if capitalism didn't invent war, it has frequently chosen it to great benefit. The United States is a good example of this tendency.

In modern times, war and the steady drumbeat for military "preparedness" have had important short- and medium-term benefits for capitalist economies, particularly in hard economic times. Here are a few:

- Maintaining high levels of spending for military production, thus helping to sustain growth, corporate profits, and jobs—economic stimulus programs in themselves.
- Projecting national economic interests—i.e., *corporate* interests—into distant regions to secure resources against competitors.
- Providing intimidation of potential economic and military adversaries—aided by hundreds of military bases on foreign soils and in faraway places. These in turn require networks of profitable commercial services to maintain the bases, from McDonald's restaurants to garden-furniture sales in PX stores.
- Arousing and uniting a domestic public toward common external enemies, thus gaining political support at home in hard times.
- Finding opportunities to usefully deploy old military stockpiles—which then have to be replenished. Keeping the inventory moving.

All of these activities—on the battleground, in forward-base locations, and on the home front—can be viewed as an alternative form of economic development. The "war economy" has advantages over more routine development models, in that its benefits usually come free of normal marketplace rules and procedures, including competitive bidding on contracts and regulations to prevent cost overruns.

While government officials usually launch weapons development programs, the details of the decisions are likely to have been made in tandem with the interests of large industries and corporations. Competitive bidding for large contracts is more the exception than the rule.

Is this really capitalism? Or is it a bit more like state capitalism (as in China or Russia) where state interests are often merged with corporate interests? In a prior context, Italy and Germany in the 1930s, a similar degree of state-corporate-military merger was called fascism. Chalmers Johnson, a former U.S. intelligence officer and author of *Blowback: The Costs and Consequences of American Empire*, labels the whole military-economic merger as "military Keynesianism"—military versions of corporate stimulus programs during stressful times.

Massive military spending for World War II was the main factor that lifted us out of the 1930s Great Depression. The war effort expanded industrial production and innovation in the 1940s, providing jobs to women as well as men and motivating public spirit as nothing has since. It was paid for by common sacrifice, from the soldiers sent abroad to consumers who endured rationing. Even the wealthiest Americans contributed greatly back then, enduring substantial tax increases without forming Tea Parties.

At the end of World War II, the world quickly found itself in a new global economic and geopolitical crisis, a realignment that split the countries of the world into communist and noncommunist blocs and produced a dangerous Cold War that would last for nearly a half-century. A new, modernized form of economic globalization was quickly created and began to be deployed among noncommunist countries. It sought to homogenize and integrate all "free world," capitalist economies in the cause of expanding and revitalizing markets through reconstruction, development, free movement of capital, and free trade.

By 1950 in the United States, the efforts at economic revival were proceeding. With the Depression still in mind—and with the new challenges of militant communism and its potential to stimulate constant skirmishes—many felt that high spending on military preparedness should be the new normal. The two factors combined—a military economy integrated with an expanding consumer economy—seemed to be a good way to resist the advancing threats of communist competition while also continuing to support the post-Depression recovery.

The Advent of Permawar

On April 14, 1950, five years after the end of World War II, the National Security Council under Paul Nitze issued its infamous NSC-68 report, advocating the formal merger of military policy with economic policy. President Truman accepted and signed the report. The United States immediately began drafting its basic strategy for the Cold War and beyond, putting our country on what has become a virtually permanent war footing. Military spending in the United States has now advanced to become roughly equal to the combined spending of all other countries in the world. But even now, a half-century later, there is very little awareness of the fact that our defense strategies have merged with our economic strategies.

In 1961, eleven years into this new policy, an apparently worried President (and former general) Dwight David Eisenhower delivered a remarkable warning against the whole trend in his farewell address. Eisenhower famously spoke of the dangers of a growing "military-industrial complex," adding that "the conjunction of an immense military establishment and a large arms industry is new in the American experience."

Of course, as president, Eisenhower himself actually contributed to the problem. His "Atoms for Peace" initiative aggressively promoted a push to persuade Japan that its fears of atomic energy were unfounded. Japan was asked to accept boatloads of new nuclear power technology from General Electric, which was, at that time, the corporate prince of the military-industrial complex. Atoms for Peace really amounted to yet another bailout strategy for military industries that had no big war to fight anymore. GE obtained many of the postwar construction contracts for nuclear energy in Japan, helping to give birth to the mess at Fukushima.

By the 1960s, the whole industrial-military-merger trend was too advanced, and too profitable, to reverse. Giant corporations have been making good livings from military contracts ever since. For some of them, the contracts have represented the great majority of their businesses.

But one wonders: How did our leaders in the White House and in Congress ever manage to justify building 32,000 nuclear bombs? As a practical matter, these could never possibly be used. You need only a handful of those bombs to blow up the world. Even now, we have about 8,500 of them, several thousand of which are "live."

Similarly, we must also ask if the overall threats we face justify spending nearly $1 trillion in FY 2017 on basic and hidden military costs. (In April 2016, the Center for Strategic and International Studies noted that adding "defense-related funding" to the basic $524 billion defense budget would bring the total tab to $905 billion.)

Since World War II there has been constant pressure to be alert to new enemies in order to justify military production at a high level. First, we had the Soviets to keep us focused during the Cold War, and then came the Chinese Maoists and the defense against potential threats to Formosa [the former name of Taiwan]. Soon after came invasions of Cuba, Korea, and Vietnam. After that came 9/11 and the "war on terror," and then we jumped into Afghanistan, Iraq, and Pakistan. The United States has also mounted invasions of Panama, the Dominican Republic, and Granada to fill in some threat gaps and produce minor military dramas, as well as "peacekeeping missions" in Bosnia and Kosovo. More recently there's been Libya and Syria, with North Korea and Iran still waiting their turn. Not to mention China.

Doing the Numbers

The scale of U.S. military spending is breathtaking. The following are a few highlights.

Between 2006 and 2011, U.S. military expenditures accounted for more than 45 percent of all discretionary spending of U.S. tax dollars—a total military expenditure that even the Pentagon publicly acknowledges as now being well over $720 billion per year. (These figures do not include "national security" spending hidden in the budgets of other U.S. departments—i.e., Energy, State, Treasury, Veterans Affairs, CIA, NSA, et al. Adding these would likely make the total more than $1 trillion per year.) Even if the oft-quoted $720 billion official figure is accurate, that's roughly *half of all the military expenditures of all countries in the world, combined.*

Under President Obama's proposed $4.2 trillion FY 2017 spending plan, $622.6 billion of the discretionary budget would flow directly to the Pentagon. Veterans programs would receive an added $75.4 billion, totaling 61 percent of the discretionary budget. The remaining 39 percent would attempt to cover people's needs—for education, transportation, housing, health, science, environmental protection, etc.

The United States is projected to spend about four times as much on the military as our next largest competitor, China, which, despite having a population roughly four times larger than ours, spends only about $146 billion annually on its military. According to the Stockholm International Peace Research Institute (SIPRI), using 2016 numbers, the top ten countries for military expenditures were:

1. United States: $597.5 billion
2. China: $145.8 billion
3. Saudi Arabia: $81.9 billion
4. Russia: $65.6 billion
5. United Kingdom: $56.2 billion
6. India: $48 billion
7. France: $46.8 billion
8. Japan: $41.0 billion
9. Germany: $36.7 billion
10. South Korea: $33.5 billion

As for our archenemies, Iran and North Korea, according to the *CIA World Factbook*, in 2008 they spent only $7.04 billion and $7 billion, respectively, on their militaries, about 1/100th of what we did.

Since the Vietnam War, total U.S. military expenditures have exceeded $20 trillion and, according to the Congressional Budget Office, U.S. Department of Defense spending between 2001 and 2010 increased an average of 9 percent per year. The usual explanations for this spectacular growth in military spending were, of course, 9/11, the al-Qaeda threat, and the "war on terror." But terrorism is not conventional warfare requiring multibillion-dollar aircraft carriers and $80 million jet fighters. Yet most of the U.S. budget has gone to this traditional weaponry—far more helpful for corporate capitalist needs than for fighting suicide bombers.

Over the last decade, a high percentage of total spending was concentrated on a small number of contractors. During 2009, for example, the top fifteen corporations receiving U.S. defense contracts accounted for more than a quarter of all U.S. military/defense spending. And although the ranking among them shifted somewhat from year to year, the majority of these corporations have been the same over the last decade, suggesting a close working relationship between these companies and the U.S. military procurement process. In FY 2015, according to the U.S. General Services Administration's Federal Procurement Data System, the top five were:

1. Lockheed Martin Corp.: $36.3 billion
2. Boeing Corp.: $16.6 billion
3. General Dynamics Corp.: $13.6 billion
4. Raytheon Co.: $13.1 billion
5. Northrop Grumman Corp.: $10.6 billion

The top fifteen included: BAE Systems Inc., Bechtel Group Inc., Booz Allen Hamilton Holding Corp., General Electric Co., L-3 Communications Corp., McKesson Corp., Science Applications International Corp., and United Technologies Corp.

A further expression of government support for aerospace/defense industry contractors is that the effective tax rate for the average defense industry corporation is 1.8 percent. For all other corporations in the United States, the average tax rate is 18.4 percent. That differential amounts to yet another hidden subsidy for militarization that doesn't get included in the overall military budget.

Christopher Hellman, of the National Priorities Project, notes a long list of war and national security expenditures that never show up in the totals reported by the Pentagon.

For example, in the FY 2012 federal budget projections, the Department of Energy received $19.3 billion for keeping nuclear weapons properly maintained and for cleaning up waste from the weapons; spying and intelligence-gathering functions of the National Intelligence Program (including assignments from the CIA and NSA) totaled $53 billion; expenses by NASA for spy satellites added $18.7 billion. Another $53.5 billion went to pay carryover expenses for past wars, and for "general support for current and future national security strategy" via various other federal accounts, including the Department of Homeland Security ($37 billion), the Department of Health and Human Services ($4.6 billion), and the Department of Justice ($4.6 billion).

It is important to remember that most of these military expenditures actually wind up in the hands of private corporations performing services in each of these categories. Another very important dimension of this story is that U.S. corporate arms manufacturers also do very big business selling to other countries.

In 2015, total U.S. arms sales to global markets (more than $46 billion according to the Defense Security Cooperation Agency) represented more than half of all global arms sales in the world, and that percentage is increasing. In October 2010, the Obama administration authorized the sale of some $60 billion in armaments to the government of Saudi Arabia, where 9/11 was born. These included Apache attack helicopters, tactical Black Hawk helicopters, and F-15 fighter jets. It was the biggest commercial armament sale in U.S. history. Not announced, but expected by most observers: an offsetting sale to Israel to assure its own military readiness would remain superior to Saudi Arabia's. Sure enough, a record $3 billion in aid to Israel was included in the 2012 budget and, in 2015, Obama approved additional sales of $1.9 billion in missiles, bombs, and rockets to Israel. All of this can only be viewed realistically as yet more economic stimulus.

Military Keynesianism

A gigantic percentage of military hardware (and software) is utterly out of date for modern "fourth-generation warfare." Much of it exists only for the purpose of subsidizing arms manufacturers. They are essentially bailouts and job stimulus programs for an economy that doesn't seem to be able to sustain jobs in other ways.

Sometimes the motives are very direct and clear, as with the production of the Lockheed Martin F-22 Stealth fighter. Lockheed Martin began construction of the F-22 in 1986, at a cost of $200 million each. Twenty-five years later, as the last F-22 emerged from Lockheed's Georgia assembly plant, *Wired* magazine reported each individual fighter had cost "as much as $678 million, depending on how and what you count." (With 183 F-22s on order, that would boost the total cost to $124 billion.)

Another case concerns the B-2 bomber, which was originally meant to replace the Fairchild A-10 "Warthog," which had performed well in the first Persian Gulf War. Northrop Grumman and the Pentagon argued that what we really needed now, rather than A-10s, were new, superfast (less maneuverable), high-altitude bombers that flew in straight lines. Three of these new Northrop Grumman B-2s cost the equivalent of 715 of the A-10s. According to Chalmers Johnson, the B-2 stealth bombers "are too delicate to deploy to harsh climates without special hangars first being built to protect them, at ridiculous expense; and they cannot fulfill any combat missions that older designs were not fully adequate to perform; and, at a total cost of $44.75 billion for only 21 bombers."

In a March 13, 2011, *New York Times* article titled "The Pentagon's Biggest Boondoggles," John Arquilla cites other examples of the most excessive and utterly useless Defense Department projects. These include:

F-35 Lightning II Fighter (Lockheed Martin Corporation):

The original estimate was that 2,443 of these planes, begun in 2000, would cost $178 billion. By 2012, the cost per plane topped $304 million. By 2014, the program was $163 billion over budget and seven years behind schedule. By 2016, the tab was approaching $400 billion, more than double the original estimate. One Pentagon report concluded: "Affordability is no longer embraced as a core pillar."

Gerald Ford-Class Supercarrier (Northrop Grumman Corporation):

The estimated cost for this giant, 100,000-ton aircraft carrier was initially budgeted at $5 billion per ship when work began in 2007 but has increased to $13 billion each. Ten ships were supposed to be ready by 2015 but (according to a December 2016 report in the *Navy Times*) the first, the *USS Gerald R. Ford* (christened on November 9, 2013), is not set to be deployed until sometime in 2021. As CNN reported on July 27, 2016, "The most expensive warship in history continues to struggle launching and recovering aircraft, moving onboard munitions, conducting air traffic control and with ship self-defense." In November 2014, the Government Accountability Office announced the ship had gone 22 percent over budget and the Navy was planning on meeting the "price cap" by accepting an unfinished version of the ship. "In essence," the GAO reported, "the Navy will have a ship that is less complete than initially planned at ship delivery, but at a greater cost."

Littoral Combat Ship (Austal USA and Lockheed Martin):

By 2015, Austal and Lockheed were supposed to deliver ten high-speed, shallow-draft, catamaran-hulled vessels intended to fight running battles in shallow coastal waters of the Pacific, presumably against China. The initial budget ($220 million for each of 55 ships) has risen to $650 million per vessel. John McCain excoriated the LCS design as "not operationally effective or reliable," noting that the aluminum superstructure "will burn to the waterline if hit."

The Atomic Arsenal

"Between the 1940s and 1996," according to Chalmers Johnson, "the U.S. spent at least $5.8 trillion on the development, testing, and construction of nuclear bombs. By 1967, the peak year of its nuclear stockpile, the U.S. possessed some 32,500 deliverable atomic and hydrogen bombs. This perfectly illustrates the Keynesian principle that the government can produce profits for giant corporations and provide make-work jobs to keep people employed. Nuclear weapons were not just America's special weapon, but its special economic weapon."

In *Dismantling the Empire*, Johnson summarizes the situation this way: "In our devotion to militarism, despite our limited resources, we are failing to invest in our social infrastructure and other requirements for the long-term health of our country. . . . Most important, we have lost our competitiveness as a manufacturer for civilian needs . . . an infinitely more efficient use of scarce resources than arms manufacturing."

Johnson goes into great detail about how U.S. military spending—often of no practical use, sometimes poorly produced, and with inflated production costs—is introduced and sustained. It's a process characterized by shameless lobbying, self-interest, political campaign debt payments, handouts, and bailouts.

Franklin "Chuck" Spinney, a former high-level analyst with the Pentagon's Office of Systems Analysis, has also criticized the irrationality of the military purchasing process. Spinney described a contracting process featuring two corrupt and distorting tendencies: (1) "front loading," in which powerful corporate-military lobbies join forces to achieve huge up-front government payments for unproven technology that often ultimately fails, and (2) "political engineering," the practice of seeking contracts in as many congressional districts as possible simultaneously to solidify congressional support for over-budgeted and frequently useless equipment.

The effectiveness of this "all voting districts" strategy was demonstrated when Congress attempted to cancel the B-2 bomber in 1990. After the disintegration of the Soviet Union, the B-2, which had been designed specifically against Russian defense systems, lost its *raison d'être*. However, the Pentagon and Northrop Grumman created a gigantic lobbying campaign that emphasized that tens of thousands of jobs and hundreds of millions in profits were at risk in 46 states and 383 congressional districts. As a result, the B-2 is still with us.

"Comparative Advantage" of War

We have thousands of fancy airplanes and ships that cost hundreds of billions of dollars but that we don't really need. We have thousands of bombs that can destroy the planet but that we can never use. We are occupying hundreds of far-flung bases all over the world, and small islands from which we have expelled populations against their will (or transformed them into colonized restaurant workers and janitors). We've redesigned the islands into militarized versions of Las Vegas. We continue to occupy these islands because of, we now say, growing threats from China. But China occupies no islands and is still gaining on us economically. Why do we still do this?

The best answer may lie in classical Ricardian economic theories of "comparative advantage," as follows:

Not every country produces everything that it might produce. For example, the United States could produce more stained-glass windows and high-quality wines instead of military hardware. But we don't have much of a history of that activity; we don't have the artists or the production mechanisms; we don't even know how to think about it.

But there is one thing the United States knows how to do: make armaments. We are a big country and are still pretty rich. (Any country involved in the armaments industry has to be very big and rich to indulge in it.) We already have a highly developed military-industrial infrastructure well in place, and have a constant market in all those bases around the world, requiring nonstop material replenishment.

In terms of capitalism, therefore, military production and marketing may actually be our best economic opportunity—we already dominate it and it seems like we can keep it growing at 9 to 10 percent per year. What's more, this market doesn't need real customers. We are our own customers. We do sell a small percent of this military production to other countries—more than the rest of the world combined—but mostly, we consume the products our companies produce, deploy them, and use them up ourselves (or leave them in storage somewhere forever). Anyway, there are always new wars to eventually use up our stockpiles, or new threats of wars to provide profitable new markets and good jobs.

So, speaking economically, we are better off concentrating on what we know and what we do best, letting other countries deploy their more advanced skills in light cars, solar panels, wind arrays, fast trains, green technology, good food, good wine, and fine inexpensive fabrics.

Military production is one economic area where we can easily dominate global supply, production, distribution, and deployment and achieve constant renewal of the process. It is American capitalist genius at its best.

An Empire of Military Bases

Hugh Gusterson (2009)

How many military bases does the United States have in other countries? According to the Pentagon's own list, the answer is around 865, but if you include the new bases in Iraq and Afghanistan, it is over a thousand. These bases constitute 95 percent of all the military bases any country in the world maintains on any other country's territory. In other words, the United States is to military bases as Heinz is to ketchup.

The old way of doing colonialism, practiced by the Europeans, was to take over entire countries and administer them. But this was clumsy. The United States has pioneered a leaner approach to global empire. As historian Chalmers Johnson says, "America's version of the colony is the military base." The United States, says Johnson, has an "empire of bases."

"Its 'empire of bases' gives the United States global reach, but the shape of this empire, insofar as it tilts toward Europe, is a bloated and anachronistic holdover from the Cold War."

These bases do not come cheap. Excluding U.S. bases in Afghanistan and Iraq, the United States spends about $102 billion a year to run its overseas bases. (In 2015, maintaining bases in Iraq and Afghanistan cost around $71 billion.) And in many cases you have to ask what purpose they serve. For example, the United States has 227 bases in Germany. Maybe this made sense during the Cold War, when Germany was divided by the Iron Curtain and U.S. policymakers sought to persuade the Soviets that the American people would see an attack on Europe as an attack on itself. But, in a new era, when Germany is reunited and the United States is concerned about flashpoints of conflict in Asia, Africa, and the Middle East, it makes as much sense for the Pentagon to hold on to its 227 German bases as it would for the U.S. Post Office to maintain a fleet of horses and buggies.

The White House is desperate to cut unnecessary costs in the federal budget. In 2008, Congressperson Barney Frank (D-MA) suggested cutting the Pentagon budget by 25 percent. In 2004, Defense Secretary Donald Rumsfeld estimated that the United States could save $12 billion by closing around 200 foreign bases.

Today, most politicians and media pundits seem oblivious to these bases, treating the stationing of U.S. troops all over the world as a natural fact. Meanwhile, the U.S. empire of bases has been attracting increasing attention from academics. In 2009, NYU Press published Catherine Lutz's *Bases of Empire*, a study of U.S. military bases and protests against the bases. Rutgers University Press published Kate McCaffrey's 2002 investigation, *Military Power and Popular Protest*, a study of the U.S. base at Vieques, Puerto Rico (closed in the face of massive protests from the local population). And in 2011, Princeton University Press published David Vine's *Island of Shame*—which tells the story of how the United States and Britain secretly agreed to deport the Chagossian inhabitants of Diego Garcia to Mauritius and the Seychelles so their island could be turned into a military base.

American leaders speak of foreign bases as cementing alliances with foreign nations, largely through the trade and aid agreements that often accompany base leases. Yet, U.S. soldiers live in a cocooned simulacrum of America in their bases, watching American TV, listening to American rap and heavy metal, and eating American fast food, so that the transplanted farm boys and street kids have little exposure to another way of life. Meanwhile, on the other side of the barbed-wire fence, local residents and businesses often become economically dependent on the soldiers. At the same time, these bases can become flashpoints for conflict. They invariably discharge toxic wastes into local ecosystems—as in Guam, where military bases are responsible for nineteen Superfund sites. Such contamination generates resentment and sometimes, as in Vieques in the 1990s, full-blown social movements against the bases. The United States used Vieques for live-bombing practice 180 days a year and, by the time the Pentagon withdrew in 2003, the landscape was littered with exploded and unexploded ordinance, depleted uranium rounds, heavy metals, oil, lubricants, solvents, and acids. According to local activists, the cancer rate on Vieques was 30 percent higher than in the rest of Puerto Rico.

It is also inevitable that, from time to time, U.S. soldiers—often drunk—commit crimes. The resentment these crimes cause is exacerbated by Washington's insistence that such crimes cannot be prosecuted in local courts.

In 2002, two U.S. soldiers killed two teenage girls in Korea as they walked to a birthday party. Korean campaigners claim this was one of 52,000 crimes committed by U.S. soldiers in Korea between 1967 and 2002. The two soldiers were immediately repatriated to the United States so they could escape prosecution in Korea.

In 1998, a marine pilot sliced through the cable of a ski gondola in Italy, killing twenty people, but U.S. officials refused to allow Italian authorities to try him. These and other similar base-related incidents have injured U.S. relations with allies.

The 9/11 attacks are arguably the most spectacular example of the kind of blowback that can be generated from local resentment against U.S. bases. In the 1990s, the presence of U.S. military bases near the holiest sites of Sunni Islam in Saudi Arabia angered Osama bin Laden and provided al-Qaeda with a potent recruitment tool. The United States wisely closed its largest bases in Saudi Arabia, but it opened additional bases in Iraq and Afghanistan that quickly became new sources of friction between the United States and the peoples of the Middle East.

At a time of record budget deficits, many of these bases are a luxury the United States can no longer afford. While U.S. foreign bases project American power across the globe, they also inflame foreign relations by generating resentment against the prostitution, environmental damage, petty crime, and everyday ethnocentrism that are their inevitable corollaries. Such resentments have forced the closure of U.S. bases in Ecuador, Puerto Rico, and Kyrgyzstan. More movements against U.S. bases can be expected in the future.

The Declaration of Independence criticizes the British "for quartering large bodies of armed troops among us" and "for protecting them, by a mock trial, from punishment for any murders, which they should commit on the inhabitants of these States."

Fine words! The United States should start taking them to heart.

The High Costs of a Warfare Economy

Dave Gilson (2014)

Until Senator Patty Murray (D-WA) and Representative Paul Ryan (R-WI) rode to the rescue in December 2013, Pentagon brass and their allies had been issuing dire warnings about the nation's military readiness: the armed services were being decimated, they said, by "sequestration"—the automatic budget cuts that were set to trim $1 trillion from the Pentagon budget over the next decade. "It's one thing for the Pentagon to go on a diet. It's another for the Pentagon to wear a straitjacket while dieting," grumbled Representative Jim Cooper (D-TN). The message got through: the House overwhelmingly approved the Ryan-Murray plan just two days after it was introduced.

The Pentagon had once more gotten a reprieve from the budget ax: under Murray and Ryan's congressional budget deal, the Pentagon was granted an additional $32 billion (or 4.4 percent) in 2014, leaving its base budget at a higher level than in 2005 and 2006.

Before the budget deal, some defense spending critics had been ready to accept sequestration as the blunt, imperfect tool that might force the military to shed some of the bulk it acquired while fighting two of the longest and most expensive wars in our history. Even with the sequester in place, the Pentagon's base budget was set to remain well above pre-9/11 levels for the next decade.

The wars in Iraq and Afghanistan cost $1.5 trillion—about twice the cost of the Vietnam War when adjusted for inflation. Those funds came entirely from borrowing, contributing nearly 20 percent to the national debt accrued between 2001 and 2012. And that's just the "supplemental" military spending passed by Congress for the wars—the regular Pentagon budget also grew nearly 45 percent between 2001 and 2010.

No wonder that defense watchdogs found the Pentagon's wailing about the sequester less than convincing. "These 'terrible' cuts would return us to historically high levels of spending," snapped Winslow Wheeler of the Project on Government Oversight. According to Lawrence J. Korb, a senior fellow at the Center for American Progress, the Pentagon could reduce its budget by $100 billion a year without undermining its readiness. The sequestration cuts for 2013 amounted to $37 billion.

Not so long ago, a hawkish GOP politician called for the "bloated" defense establishment to "be pared down" and retooled for the twenty-first century. (That GOP politician? Former senator Chuck Hagel, who was appointed secretary of defense in 2013.) Below, a field guide to just how big the Pentagon budget is—and why it's so hard to trim.

Our Military Is Mind-Bogglingly Big

- The Pentagon employs three million people, 800,000 more than Walmart.
- The Pentagon's 2012 budget was 47 percent bigger than Walmart's.
- Serving 9.6 million people, the Pentagon and Veterans Administration together constitute the nation's largest health-care provider.

- Seventy percent of the value of the federal government's $1.8 trillion in property, land, and equipment belongs to the Pentagon.
- Los Angeles could fit into the land managed by the Pentagon 93 times.
- The Army uses more than twice as much building space as all the offices in New York City.
- The Pentagon holds more than 80 percent of the federal government's inventories, including $6.8 billion of excess, obsolete, or unserviceable stuff.
- One out of every five tax dollars is spent on defense.

The $3.7 trillion federal budget breaks down into mandatory spending—benefits guaranteed the American people, such as Social Security and Medicare—and discretionary spending—programs that, at least in theory, can be cut. In 2013, more than half of all discretionary spending (and one-fifth of total spending) went to defense, including the Pentagon, veterans' benefits, and the nuclear weapons arsenal.

We're still the world's 800-pound gorilla. When it comes to defense spending, no country can compete directly with the United States, which spends more than the next ten countries combined—including potential rivals Russia and China, as well as allies such as England, Japan, and France. The Pentagon accounts for nearly 40 percent of global military spending. In 2012, 4.4 percent of our GDP went to defense. That's in line with how much Russia spends. China spends 2 percent of its GDP on its military.

Where does the Pentagon's money go? The exact answer is a mystery. That's because the Pentagon's books are a complete mess. They're so bad that they can't even be officially inspected, despite a 1997 requirement that federal agencies submit to annual audits—just like every other business or organization.

The Defense Department is one of just two agencies (Homeland Security is the other) that are keeping the bean counters waiting: as the Government Accountability Office (GAO) dryly notes, the Pentagon has "serious financial management problems" that make its financial statements "inauditable." Pentagon financial operations occupy one-fifth of the GAO's list of federal programs with a high risk of waste, fraud, or inefficiency.

Critics also contend that the Pentagon cooks its books by using unorthodox accounting methods that make its budgetary needs seem more urgent. The agency insists it will "achieve audit readiness" by 2017.

A Long History of Overruns and Rip-Offs

1778: Gen. George Washington decries the suppliers overcharging his army: "It is enough to make one curse their own Species for possessing so little virtue & patriotism."

1861: A House committee exposes fraud, favoritism, and profiteering in Civil War contracting. Its findings, writes the *New York Times*, "produce a feeling of public indignation which would justify the most summary measures against the knaves whose villainy is here dragged into daylight."

1941: Senator Harry S. Truman kicks off a dogged investigation of wasteful war production. The Truman Committee, which runs into World War II, is credited with saving as much as $15 billion ($245 billion in 2017 dollars).

1975: Wisconsin senator William Proxmire calls out the Pentagon for maintaining 300 golf courses around the world. (Today it has at least 170.)

1983: President Ronald Reagan announces the Strategic Defense Initiative, aka "Star Wars," a system of ground-and-space-based lasers designed to stop incoming nuclear missiles. Still unrealized, the program has cost more than $209 billion.

1985: Pentagon profligacy makes headlines with reports of $640 toilet seats, $660 ashtrays, $7,600 coffeemakers, $74,000 ladders, and a $436 hammer.

2001: Halliburton subsidiary KBR takes over a contract to feed soldiers in Iraq. It raises the price of a meal from $3 to $5 while subcontracting the services back to the previous contractor.

2009: After safety problems and cost overruns, the Pentagon cancels the F-22 Raptor fighter jet (estimated price tag: $412 million per plane).

2010: The GAO finds that the Defense Logistics Agency is sitting on $7.1 billion worth of excess spare parts.

2010: An anonymous congressional earmark sets aside $2.5 billion for ten C-17 aircraft the Air Force says it does not need.

2011: Boeing charges the Army $1,678 apiece for rubber cargo-loading rollers that actually cost $7 each.

2012: One-quarter of the $1.6 trillion being spent on major weapons systems comes from unexpected cost overruns.

2013: The Pentagon plans to scrap more than 85,000 tons of equipment in Afghanistan, part of $7 billion worth of gear being left behind as the troops come home.

Anatomy of a Budget-Buster

In the early 2000s, the Pentagon began developing the F-35 Joint Strike Fighter, a new generation of stealthy, high-tech jets that were supposed to do everything from landing on aircraft carriers and taking off vertically to dog-fighting and dropping bombs. The F-35 has earned a reputation as the biggest defense boondoggle in history. Rolling out the F-35 is expected to cost nearly $400 billion. The time needed to develop the plane has gone from ten years to eighteen. According to the GAO, the actual cost has jumped from $75 million per plane to $137 million.

Planes started rolling off the assembly line before development and testing were finished, which could result in $8 billion worth of retrofits. A 2013 report by the Pentagon inspector general identified 719 problems with the F-35, including:

- Pilots are not allowed to fly these test planes at night, within twenty-five miles of lightning, faster than the speed of sound, or with real or simulated weapons.
- Pilots say cockpit visibility is worse than in existing fighters.
- Special high-tech helmets have "frequent problems" and are "badly performing."
- Takeoffs may be postponed when the temperature is below 60°F.

So why does this boondoggle persist? Perhaps it's because the F-35 program has 1,400 suppliers in 46 states. Lockheed Martin gave money to 425 members of Congress in 2012 and spent $159 million on lobbying between 2000 and 2014.

Guns and Butter

A close look at the $361 billion handed to military contractors in 2012 reveals the enormous amount of stuff the modern military consumes. Some of the items on the shopping list:

- Planes and helicopters: $32.5 billion
- Petroleum and oil: $21.6 billion
- Guided missiles: $10.4 billion
- Combat/assault vehicles: $5.2 billion
- Dairy and eggs: $4 billion
- Amphibious assault ships: $3.9 billion
- Space vehicles: $3.6 billion
- Submarines: $3.4 billion
- Nuclear reactors: $2.5 billion
- Drugs and pharmaceuticals: $2.5 billion
- Combat ships and landing vessels: $2.2 billion
- Unmanned aircraft and drones: $2.2 billion
- Aircraft carriers: $1.5 billion
- Meat, poultry, and fish: $1.2 billion
- Bombs: $1 billion
- Small-arms ammunition: $978 million
- Naval destroyers: $977 million
- Night-vision equipment: $834 million
- Fruits and vegetables: $783 million
- Bakery and cereal products: $738 million
- Nonalcoholic beverages: $554 million
- Land mines: $547 million
- Small arms: $423 million
- Sugar, confectionery, and nuts: $294 million.

Cutting the Pork

Here are ten ideas for major cuts (and estimated savings) from the libertarian Cato Institute and the liberal Center for American Progress:

- Get rid of all ICBMs and nuclear bombers ($20 billion/year).
- Retire two of the Navy's 11 aircraft carrier groups ($50 billion through 2020).
- Cut the size of the Army and Marines to pre-9/11 levels (at least $80 billion over 10 years).
- Slow down or cancel the F-35 fighter jet program (at least $4 billion/year).
- Downsize military headquarters that grew after 9/11 ($8 billion/year).
- Cancel the troubled V-22 Osprey tilt-rotor (at least $1.2 billion).
- Modify supplemental Medicare benefits for veterans ($40 billion over 10 years).
- Scale back purchases of Littoral Combat Ships ($2 billion in 2013).
- Cap spending on military contractors below 2012 levels ($2.9 billion/year).
- Retire the Cold War–era B-1 bomber ($3.7 billion over 5 years).

Disaster Capitalism on the Battlefield

William J. Astore (2013)

There is a new normal in America: our government may shut down, but our wars continue. Congress may not be able to pass a budget, but the U.S. military can still launch commando raids in Libya and Somalia, the Afghan War can still be prosecuted, Italy can be garrisoned by American troops (putting the "empire" back in Rome), Africa can be used as an imperial playground (as in the late nineteenth-century "scramble for Africa," but with the United States and China doing the scrambling this time around), and the military-industrial complex can still dominate the world's arms trade.

In the halls of Congress and the Pentagon, it's business as usual, if your definition of "business" is the power and profits you get from constantly preparing for and prosecuting wars around the world.

Once upon a time, as a serving officer in the U.S. Air Force, I was taught that Carl von Clausewitz had defined war as a continuation of politics by other means. This definition is, in fact, a simplification of his classic and complex book, *On War*, written after his experiences fighting Napoleon in the early nineteenth century.

The idea of war as a continuation of politics is both moderately interesting and dangerously misleading: interesting because it connects wars to political processes and suggests that they should be fought for political goals; misleading because it suggests that war is essentially rational and so controllable. The fault here is not Clausewitz's, but the American military's for misreading and oversimplifying him.

Perhaps another "Carl" might lend a hand when it comes to helping Americans understand what war is really all about. I'm referring to Karl Marx, who admired Clausewitz, notably for his idea that combat is to war what a cash payment is to commerce. However seldom combat (or such payments) may happen, they are the culmination and so the ultimate arbiters of the process.

War, in other words, is settled by killing, a bloody transaction that echoes the exploitative exchanges of capitalism. Marx found this idea to be both suggestive and pregnant with meaning. So should we all. Following Marx, Americans ought to think about war not just as an extreme exercise of politics, but also as a continuation of exploitative commerce by other means. Combat as commerce.

In the history of war, such commercial transactions took many forms, whether as territory conquered, spoils carted away, raw materials appropriated, or market share gained. Consider American wars. The War of 1812 is sometimes portrayed as a minor dust-up with Britain, but it really was about crushing Indians on the frontier and grabbing their land. The Mexican-American War was another land grab, this time for the benefit of slaveholders. The Spanish-American War was a land grab for those seeking an American empire overseas, while World War I was for making the world "safe for democracy"—and for American business interests globally.

Even World War II, a war necessary to stop Hitler and Imperial Japan, witnessed the emergence of the United States as the arsenal of democracy, the world's dominant power, and the new imperial stand-in for a bankrupt British Empire.

Korea? Vietnam? Lots of profit for the military-industrial complex and plenty of power for the Pentagon establishment. Iraq, the Middle East, current adventures in Africa? Oil, markets, natural resources, global dominance.

In societal calamities like war, there will always be winners and losers. But the clearest winners are often companies like Boeing and Dow Chemical, which provided B-52 bombers and Agent Orange, respectively, to the U.S. military in Vietnam. Such "arms merchants"—an older, more honest term than today's "defense contractor"—don't have to pursue the hard sell, not when war and preparations for it have become so permanently, inseparably intertwined with the American economy, foreign policy, and our nation's identity as a rugged land of "warriors" and "heroes."

Consider one more definition of war: not as politics or even as commerce, but as societal catastrophe. Thinking this way, we can apply Naomi Klein's concepts of the "shock doctrine" and "disaster capitalism" to it. When such disasters occur, there are always those who seek to turn a profit.

Most Americans are, however, discouraged from thinking about war this way, thanks to the power of what we call "patriotism." During wars, we're told to "support our troops," to wave the flag, to put country first, to respect the patriotic ideal of selfless service and redemptive sacrifice (even if all but 1 percent of us are never expected to serve or sacrifice).

We're discouraged from reflecting on the uncomfortable fact that as "our" troops sacrifice and suffer, others in society are profiting big-time. Such thoughts are considered unseemly and unpatriotic. After all, any price is worth paying (or profits worth offering up) to contain the enemy—not so long ago, the Red Menace, but in the twenty-first century, the murderous terrorist.

Forever war is forever profitable. Think of the Lockheed Martins of the world. In their commerce with the Pentagon, as well as the militaries of other nations, they ultimately seek cash payment for their weapons and a world in which such weaponry will be eternally needed. In the pursuit of security or victory, political leaders willingly pay their price.

In the Roaring Twenties, President Calvin Coolidge declared, "The business of America is business." Almost a century later, the business of America is war, even if today's presidents are too polite to mention that the business is booming.

Today's young recruits are hailed as warriors and warfighters, as heroes, and not just within the military either, but by society at large. Yet in joining the military, our troops paradoxically become yet another commodity, another consumable of the state. Indeed, they become consumed by war and its violence. Their compensation? To be packaged and marketed as the heroes of our militarized moment.

Such "heroic" identities, tied so closely to violence in war, often prove poorly suited to peacetime settings. Frustration and demoralization devolve into domestic violence and suicide. In an American society with ever fewer meaningful peacetime jobs, exhibiting greater and greater polarization of wealth and opportunity, the decisions of some veterans to turn to or return to mind-numbing drugs of various sorts and soul-stirring violence are tragically predictable.

As Russian revolutionary Leon Trotsky pithily observed, "You may not be interested in war, but war is interested in you." If war is combat and commerce, calamity and commodity, it cannot be left to our political leaders alone—and certainly not to our generals. When it comes to war, however far from it we may seem to be, we're all in our own ways customers and consumers. Some pay a high price. Many pay a little. A few gain a lot. Keep an eye on those few and you'll end up with a keener appreciation of what war is actually all about. Just remember: in the grand bargain that is war, it's their product and their profit. And that's no bargain for America—or, for that matter, for the world.

"False Flags": How Wars Are Packaged and Sold

James Corbett (2016)

In naval warfare, a "false flag" refers to an attack where a vessel flies a flag other than its true battle flag before engaging its enemy. It is a trick, designed to deceive the enemy about the true nature and origin of an attack.

In the democratic era, where governments require at least a plausible pretext before sending their nation to war, it has been adapted as a psychological warfare tactic to deceive a government's own population into believing that an enemy nation has attacked them.

In the 1780s, Swedish king Gustav III was looking for a way to unite an increasingly divided nation and raise his own falling political fortunes. Deciding that a war with Russia would be a sufficient distraction but lacking the political authority to send the nation to war unilaterally, he arranged for the head tailor of the Swedish Opera House to sew some Russian military uniforms. Swedish troops were then dressed in the uniforms and sent to attack Sweden's own Finnish border post along the Russian border. The citizens in Stockholm, believing it to be a genuine Russian attack, were suitably outraged, and the Swedish-Russian War of 1788–90 began.

In 1931, Japan was looking for a pretext to invade Manchuria. On September 18 of that year, a lieutenant in the Imperial Japanese Army detonated a small amount of TNT along a Japanese-owned railway in the Manchurian city of Mukden. The act was blamed on Chinese dissidents and used to justify the occupation of Manchuria just six months later. When the deception was later exposed, Japan was diplomatically shunned and forced to withdraw from the League of Nations.

In 1939, Heinrich Himmler masterminded a plan to convince the public that Germany was the victim of Polish aggression in order to justify the invasion of Poland. It culminated in an attack on Sender Gleiwitz, a German radio station near the Polish border, by Polish prisoners who were dressed up in Polish military uniforms, shot dead, and left at the station. The Germans then broadcast an anti-German message in Polish from the station, pretended that it had come from a Polish military unit that had attacked Sender Gleiwitz, and presented the dead bodies as evidence of the attack. Hitler invaded Poland immediately thereafter, starting World War II.

In 1954, the Israelis hired a number of Egyptian Jews to plant bombs in American and British cinemas, libraries, and other civilian targets to be blamed on the Muslim Brotherhood or other malcontents. The plan, known as the Lavon Affair, was part of an effort to convince the British to retain their military presence in the occupied Suez Canal zone. Several bombings took place, but the British were ultimately forced out after Egyptian president Gamal Abdel Nasser nationalized the canal in 1956.

In 1962, the U.S. Joint Chiefs of Staff authored a document called Operation Northwoods calling for the U.S. government to stage a series of fake attacks, including the shooting down of military or civilian U.S. aircraft, the destruction of a U.S. ship, sniper attacks in Washington, and other atrocities, to blame on the Cubans as an excuse for launching an invasion. President Kennedy refused to sign off on the plan and was killed in Dallas the next year.

In August 1964, the *USS Maddox*, a U.S. destroyer on patrol in the Gulf of Tonkin, reported coming under attack from North Vietnamese navy torpedo boats, engaging in evasive action and returning fire. The incident led to the Gulf of Tonkin resolution authorizing President Johnson to begin open warfare in Vietnam. It was later admitted that no attack occurred. In 2005, it was revealed that the National Security Agency manipulated information to make it look like an attack had taken place.

In June 1967, Israeli forces attacked the *USS Liberty*, a U.S. Navy technical research ship, off the coast of Egypt. The ship was strafed relentlessly for hours in an apparent attempt to blame the attack on Egypt and draw the Americans into the Six-Day War. NSA intercepts subsequently confirmed that the Israelis knew they were attacking an American ship, not an Egyptian ship as their cover story maintained.

In the fall of 1999, a wave of bloody apartment bombings swept through Russian cities, killing 293 people and causing widespread panic. Although blamed on the Chechen terrorists that the Russians were fighting in the Second Chechen War, Russian Federal Security Service (FSB) agents were caught planting the exact same type of bombs as in the other blasts later that month. The government claimed that the bombs were part of a security exercise and Vladimir Putin came to power as the next Russian president on the back of the terror wave later that year.

In 2001, attacks in New York and Washington were blamed on al-Qaeda as a pretext for invading Afghanistan. In the months leading up to the event, American negotiators had warned Afghanistan's Taliban that they were interested in securing right-of-way for proposed oil pipeline projects and the United States threatened it would achieve this with either a "carpet of gold" or a "carpet of bombs."

The Bush administration's first major national security directive, NSPD-9, a full-scale battle plan for the invasion of Afghanistan (including command and control, air and ground forces, and logistics) was drafted and sitting on the president's desk to be signed off on September 4, 2001, seven days before the 9/11 attacks. The invasion proceeded as planned in October.

These are but a few of the hundreds of such incidents that have been staged over the centuries to blame political enemies for attacks that they did not commit. The tactic remains in common use today, and will continue to be employed as long as populations still blindly believe whatever their governments tell them about the origins of spectacular terror incidents.

Banking on the Bomb: Investing in Nuclear Weapons

Susi Snyder and Wilbert van der Zeijden, PAX and the International Campaign to Abolish Nuclear Weapons (ICAN) (2016)

The world's nuclear-armed nations spend more than $100 billion on their nuclear forces every year. This money goes toward assembling new warheads, modernizing old ones, and building missiles, launchers, and the supporting technology to use them. While the majority of that comes from taxpayers in the nuclear-armed countries, private sector investors from many non-nuclear-armed countries also provide financing that enables the production, maintenance, and modernization of nuclear arsenals.

Declassified information and new publications (including the Royal Institute of International Affairs' 2014 book, *Too Close for Comfort*) have shown how close the world has come to using nuclear weapons. At the same time, a greater international focus has been brought on the devastating humanitarian consequences of these weapons. The stigma of continued possession of nuclear weapons is growing, as they are increasingly seen as illegitimate components of any national arsenal.

But this growing awareness has not deterred commercial interests from investing in the companies that make these doomsday weapons. Since January 2011, 411 different investors have made an estimated $402 billion available to the nuclear weapons industry. Since our 2013 *Don't Bank on the Bomb* joint PAX-ICAN report, investments in companies involved in the production of nuclear weapons have increased by nearly $89 billion, and there are an additional 114 investors.*

At this point, there is still a lack of official information in the public domain about the use, production, transfer, and stockpiling of nuclear weapons, or about investments in companies that produce these weapons. Our selection of financial institutions is limited by the fact that our report uses a threshold for practical purposes. Only share and bond holdings larger than 0.5 percent of the total number of outstanding shares of one or more of the nuclear-weapons-producing companies are listed.

28 Bomb Builders; 411 Bomb Bankers

Looking at the period starting January 2011, 411 banks, insurance companies, pension funds, and asset managers from 30 countries were found to have invested significantly in the nuclear weapon industry. A total of 254 were based in North America, 94 were based in Europe, 47 in Asia, 10 in the Middle East, 5 in the Pacific, and 1 in Africa. North American investments topped $268 billion—72 percent of the total.

In 2014, we identified 28 companies involved in the production, maintenance, and modernization of nuclear weapons. The companies were based in France, Germany, India, The Netherlands, the United Kingdom, and the United States.

The top nuclear weapons builders were: Boeing, Honeywell International, Lockheed Martin, Northrop Grumman, Airbus Group, BWX Technology INC (formerly BAE Systems), Orbital ATK, and Raytheon.

The more than $402 billion that investors made available to the nuclear-weapon-producing companies between January 2011 and September 2014 were in the form of share and bond issuances, owned or managed shares, and bonds or loans. The top 10 investors provided more than $175 billion. With the exception of BNP Paribas (France), all top ten investors were based in the United States. The top three—State Street, Capital Group, and Blackrock—invested more than $80 billion combined.

In Europe, the leading nuclear weapons investors were BNP Paribas (France), Royal Bank of Scotland (UK), and Barclays (UK). In Asia, the biggest investors were Mitsubishi UFJ Financial (Japan), Sumitomo Mitsui Financial (Japan), and the Life Insurance Corporation of India.

Challenging the Nuclear Weapon Industry

Words alone cannot prevent a nuclear catastrophe from happening. All of the good intentions and plans and meetings of diplomats or heads of state and governments will not guarantee that nuclear weapons are never used again. The only way to do that is to outlaw and eliminate nuclear weapons once and for all. Stigmatizing these inhumane and indiscriminate weapons will help. Divestment makes it clear to companies that, as long as they are associated with nuclear weapons programs, they will be considered a bad investment. Divestment is usually the result of ordinary people insisting investments with their money represent their own moral or ethical standards.

Over the past few years, divestment efforts around the nuclear weapons industry have revitalized the issue in the press and for the public, leading to more financial institutions enacting policies to ban investments in nuclear weapons producers. The ultimate aim is to get these companies to withdraw from the nuclear weapons industry completely.

Don't Bank on the Bomb 2014 profiled financial institutions that have adopted policies that prevent any financial involvement in nuclear-weapon-producing companies. Eight institutions had a public policy that was comprehensive in scope and application. These institutions—a mix of ethical banks, government-managed funds, and private institutions—were based in Italy, The Netherlands, Norway, and Sweden.

Publicizing policies that ban investment in the nuclear weapons industry can generate a ripple effect on other financial institutions. Divestment by even a few institutions based on the same ethical objection can have a significant impact on a company's strategic direction.

Governments also are starting to raise the issue of weapons investments. During sessions of the 2013 UN General Assembly Open-Ended Working Group on Nuclear Disarmament, the issue of divestment was raised and participants "discussed the role of the public and private sectors."

During the 2014 Nuclear Non-Proliferation Treaty Preparatory Committee meeting, quite a few states raised concerns about the billions of dollars allocated toward "modernization" of nuclear arsenals—especially at a time of significantly shrinking national budgets. Brazil's ambassador, Pedro Motta Pinto Coelho, said it clearly: "Nuclear disarmament is also a socioeconomic imperative. . . . It is estimated that half the amount annually invested in nuclear arsenals would be enough to achieve internationally agreed development goals on poverty reduction, including the Millennium Development Goals by 2015."

It is time to start negotiations on a treaty to outlaw the use, development, production, stockpiling, transfer, acquisition, deployment, and financing of nuclear weapons, as well as assistance with these acts under any circumstances.

Experience from the Cluster Munitions Convention shows the importance of understanding why outlawing assistance is important. Nine countries have national legislation specifically prohibiting investment in cluster munitions producers, while another twenty-seven countries have national statements clearly interpreting the prohibition on assistance to include financing.

States in the world's four existing nuclear-weapons-free zones—in Latin America, Africa, Southeast Asia, and the South Pacific—could declare similar understandings from their own free-zone treaties.

Additionally, some nations have enacted domestic legislation prohibiting companies from facilitating the manufacture of nuclear weapons. For example, in Australia it is a crime for a person or company to commit "any act or thing to facilitate the manufacture, production, acquisition or testing" of nuclear weapons anywhere in the world. There is similar legislation in New Zealand. In both countries, companies also are prohibited from providing services, including lending money, to another company if it "believes or suspects, on reasonable grounds, that the services will or may assist a weapons of mass destruction program."

For the first time in decades, we have a real opportunity to significantly advance the nuclear disarmament agenda through the development of a treaty banning nuclear weapons. Additional stigmatization by prohibiting investments will help to generate public awareness and interest in outlawing and eliminating nuclear weapons once and for all.

** PAX strives to achieve the highest level of accuracy in reporting. However, at this point, there is still a marked lack of official information available in the public domain about the use, production, transfer, and stockpiling of nuclear weapons, as well as about investments in companies that produce nuclear weapons. This essay therefore reflects official information available in the public domain known to PAX. www.paxforpeace.nl*

Recruiting America's Child Soldiers

Ann Jones (2013)

Congress surely meant to do the right thing when, in the fall of 2008, it passed the Child Soldiers Prevention Act (CSPA). The law was designed to protect kids worldwide from being forced to fight the wars of Big Men. From then on, any country that coerced children into becoming soldiers was supposed to lose all U.S. military aid.

It turned out, however, that Congress—in its rare moment of concern for the next generation—had it all wrong. In its greater wisdom, the White House found countries like Chad and Yemen so vital to the national interest of the United States that it preferred to overlook what happened to the children in their midst.

As required by CSPA, in 2013 the State Department once again listed ten countries that use child soldiers: Burma (Myanmar), the Central African Republic, Chad, the Democratic Republic of the Congo, Rwanda, Somalia, South Sudan, Sudan, Syria, and Yemen. Seven of them were scheduled to receive millions of dollars in U.S. military aid as well as what's called "U.S. Foreign Military Financing." That's a shell game aimed at supporting the Pentagon and American weapons makers by handing millions of taxpayer dollars over to such dodgy "allies" who must then turn around and buy "services" from the Pentagon or "matériel" from the usual merchants of death. You know the crowd: Lockheed Martin, McDonnell Douglas, Northrop Grumman, and so on.

Here was a chance for Washington to teach a set of countries to cherish their young people, not lead them to the slaughter. But in October, as it has done every year since CSPA became law, the White House again granted whole or partial "waivers" to five countries on the State Department's "do not aid" list: Chad, South Sudan, Yemen, the Democratic Republic of the Congo, and Somalia.

Too bad for the young—and the future—of those countries. But look at it this way: Why should Washington help the children of Sudan or Yemen escape war when it spares no expense right here at home to press our own impressionable, idealistic, ambitious American kids into military "service"?

It should be no secret that the United States has the biggest, most efficiently organized, most effective system for recruiting child soldiers in the world. With uncharacteristic modesty, however, the Pentagon doesn't call it that. Its term is "youth development program."

The Youth Deployment Program

Pushed by multiple high-powered, highly paid public relations and advertising firms under contract to the Department of Defense, its most important public face is the Junior Reserve Officers Training Corps or JROTC. What makes this child-soldier recruiting program so striking is that the Pentagon carries it out in plain sight in hundreds of private, military, and public high schools across the United States.

Unlike the notorious West African warlords Foday Sankoh and Charles Taylor (both brought before international tribunals on charges of war crimes), the Pentagon doesn't actually kidnap children and drag them bodily into battle. It seeks instead to make its young "cadets" what John Stuart Mill once termed "willing slaves," so taken in by the master's script that they accept their parts with a gusto that passes for personal choice. To that end, JROTC works on their not yet fully developed minds, instilling what the program's textbooks call "patriotism" and "leadership," as well as a reflexive attention to authoritarian commands.

The scheme is much more sophisticated than any ever devised in Liberia or Sierra Leone, and it works. The result is the same: kids get swept into soldiering—a job they will not be free to leave and, in the course of which, they may be forced to commit spirit-breaking atrocities. When they start to complain or crack under pressure (in the United States as in West Africa), out come the drugs.

The JROTC program, still spreading in high schools across the country, costs U.S. taxpayers hundreds of millions of dollars annually. It has cost an unknown number of taxpayers their children.

The Acne-and-Braces Brigades

A few years ago, at a Veterans Day parade in Boston, I wandered among the uniformed groups taking their places along the Boston Common. There were some old-timers sporting the banners of their American Legion posts, a few high school bands, and some sharp young men in smart dress uniforms: greater Boston's military recruiters. Then there were the kids. The acne-and-braces brigades—fourteen- and fifteen-year-olds in military uniforms—carrying rifles against their shoulders. Some of the girl groups sported snazzy white gloves. Far too many such groups, with far too many underage kids, stretched the length of Boston Common. They represented all branches of the military and many different local communities: African Americans, Hispanics, immigrants from Vietnam and other points South.

I asked a fourteen-year-old boy why he had joined JROTC. He wore a junior Army uniform and toted a rifle nearly as big as himself. He said, "My dad, he left us, and my mom, she works two jobs, and when she gets home, well, she's not big on structure. But they told us at school you gotta have a lot of structure if you want to get somewhere. So I guess you could say I joined up for that."

A group of girls, all Army JROTC members, told me they took classes with the boys but had their own all-girl (all–African American) drill team that competed against others as far away as New Jersey. They showed me their medals and invited me to their high school to see their trophies. They, too, were fourteen or fifteen. One said, "I never got no prizes before."

Their excitement took me back. When I was their age, growing up in the Midwest, I rose before daybreak to march around a football field and practice close formation maneuvers in the dark before the school day began. Nothing would have kept me from that "structure," that "drill," that "team," but I was in a marching band and the weapon I carried was a clarinet. JROTC has entrapped that eternal youthful yearning to be part of something bigger and more important than one's own pitiful, neglected, acne-spattered self. JROTC captures youthful idealism and ambition, twists it, trains it, arms it, and sets it on the path to war.

A Little History

The U.S. Army Junior Reserve Officers' Training Corps was conceived as part of the National Defense Act of 1916 in the midst of World War I. In the aftermath of that war, however, only six high schools took up the military's offer of equipment and instructors. A senior version of ROTC was made compulsory on many state college and university campuses, despite the then-controversial question of whether the government could compel students to take military training.

In 1973, the Nixon administration discarded the draft in favor of a standing professional "all volunteer" military force. But where were those professionals to be found? And how exactly were they to be persuaded to "volunteer"? Since World War II, ROTC programs at institutions of higher education had provided about 60 percent of commissioned officers. But an army needs foot soldiers.

Officially, the Pentagon claims that JROTC is not a recruiting program. Privately, it never considered it to be anything else. Army JROTC now describes itself as having "evolved from a source of enlisted recruits and officer candidates to a citizenship program devoted to the moral, physical, and educational uplift of American youth." Yet former defense secretary William Cohen, testifying before the House Armed Services Committee in 2000, named JROTC "one of the best recruiting devices that we could have."

With that mission in hand, the Pentagon pushed for the establishment of 3,500 JROTC units to "uplift" students in high schools nationwide. The plan was to expand into "educationally and economically deprived areas." The shoddy schools of the inner cities, the Rust Belt, the Deep South, and Texas became rich hunting grounds. All together, Army, Air Force, Navy, and Marine JROTC units now flourish in 3,402 high schools nationwide—65 percent of them in the South—with a total enrollment of 557,129 kids.

Getting with the Program

Here's how it works. The Department of Defense spends several hundred million dollars—$365 million in 2013—to provide uniforms, Pentagon-approved textbooks, and equipment to JROTC, as well as part of the instructors' salaries. Those instructors, assigned by the military (not the schools), are retired officers. They continue to collect federal retirement pay, even though the schools are required to cover their salaries at levels they would receive on active duty. The military then reimburses the school for about half of that hefty pay, but the school is still out a bundle.

An American Friends Service Committee (AFSC) report, *Trading Books for Soldiers: The True Cost of JROTC*, revealed the cost of JROTC programs to local school districts was "in some cases more than double the cost claimed by the Department of Defense," with local school districts shelling out "more than $222 million in personnel costs alone."

Public schools offering JROTC programs actually subsidize the Pentagon's recruitment drive. In fact, a JROTC class costs schools (and taxpayers) significantly more than a regular physical education or American history course.

Local schools have no control over the Pentagon's prescribed JROTC curricula, which are inherently biased toward militarism. The AFSC, Veterans for Peace, and other civic groups have compiled evidence that these classes are not only more costly than regular school courses, but also inferior in quality.

For one thing, neither the texts nor the instructors teach the sort of critical thinking central to the best school curricula today. Instead, they inculcate obedience to authority, inspire fear of "enemies," and advance the primacy of military might in American foreign policy.

Civic groups have raised other objections to JROTC, ranging from discriminatory practices—against gays, immigrants, and Muslims, for example—to dangerous ones, such as bringing guns into schools. Some units even set up shooting ranges where automatic rifles and live ammunition are used. JROTC embellishes the dangerous mystique of such weapons, making them objects to covet, embrace, and jump at the chance to use.

School Days

In one Boston inner-city school, predominantly Black, I sat in on JROTC classes where kids watched endless films of soldiers on parade, then had a go at it themselves in the school gym, rifles in hand. (I have to admit that they could march far better than squads of the Afghan National Army, which I've also observed, but is that something to be proud of?) Students also had lots of time to chat with the Army recruiter, whose desk was conveniently located in the JROTC classroom.

They chatted with me, too. A sixteen-year-old African American girl, who was first in her class and had already signed up for the Army, told me she would make the military her career. Her instructor—a colonel she regarded as the father she never knew—had led the class to believe that "our war" would go on for a very long time, or as he put it, "until we've killed every last Muslim on Earth." She wanted to help save America by devoting her life to that "big job ahead."

Stunned, I blurted out, "But what about Malcolm X?" He grew up in Boston; a boulevard not far from the school was named in his honor. "Wasn't he a Muslim?" I asked.

"Oh no, ma'am," she said. "Malcolm X was an American."

A senior boy, who had also signed up with the recruiter, wanted to escape the violence of city streets. He joined up shortly after one of his best friends, caught in the crossfire of somebody else's fight, was killed in a convenience store just down the block from the school. He told me, "I've got no future here. I might as well be in Afghanistan." He thought his chances of survival would be better there, but he worried that he had to finish high school before reporting for "duty." He said, "I just hope I can make it to the war."

What kind of school system gives boys and girls such "choices"? What kind of country?

next page: US fighter jets patrol Kuwait during Operation Desert Storm, 1991. Photo: Wikipedia.

Part II

TERRACIDE—
The WAR *on* NATURE

We know that the White Man does not understand our way of life. . . .
The Earth is not his brother but his enemy and
when he has conquered it, he moves on. . . .
He treats his Mother the Earth and his Brother the Sky like merchandise.
His hunger will eat the Earth bare and leave only a desert.

—Chief Seattle, Suquamish Nation, 1855

Our ancestors once coexisted with the natural world. Over the centuries, however, our human tribes learned to *compete* with nature, brandishing ever-sturdier tools and sharper weapons to wage increasingly savage wars—for land, water, crops, forests, minerals, and buried fuels.

The planet's biodiversity has been targeted on two fronts. On the one hand, there is an overt war mounted by tanks, warships, and supersonic bombers. On the other, there is a covert war driven by industry, whose weapons include bulldozers, logging machines, deep-sea drilling rigs, and explosives designed to blow the tops off coal-rich mountains.

Armies commanded by generals target the Earth's land and seas for control. At the same time, the armies of capital stalk the globe in search of plunder and profit. Frequently, these two armies join forces—the generals competing to seize portions of the biosphere that can be exploited and "developed" to enrich politically powerful corporations. Nature is besieged by the Military-Industrial Complex.

Industrial societies expand at the expense of nature. Cities rise and sprawl as forests and wetlands shrink. Biodiverse wildlands are transformed into chemically addicted monocultures. Oil burned to warm and cool homes, to propel aircraft, vessels, and vehicles, is exhausted into the atmosphere. As the air and oceans grow hotter, submerged nations of primordial coral blacken into extinction and polar ice fields—once stable, massive, and magnificent—shrink, crumble, and collapse.

Today, the destruction of local biomes and global systems by rapacious industry and destructive warfare has reached a tipping point. Growth-at-all-cost economies are not sustainable on a planet with finite resources.

Climate change has created an unprecedented crisis of "environmental refugees," with drought-caused crop failures and the deaths of livestock contributing to the unrest behind both the surge of Arab Spring uprisings and the growth of Nigeria's Boko Haram insurgency. A December 2015 report in the *Proceedings of the National Academy of Sciences* similarly traced the outbreak of the 2011 civil war in Syria to an extreme drought that ravaged the country from 2006 to 2009 and drove as many as 1.5 million desperate people into Syria's crowded cities.

The year 2016 was the hottest on record, according to NASA's Goddard Institute for Space Studies. (A troubling statistic, given that each of the two previous years had also been the "hottest on record.") At the end of 2016, Arctic sea ice had retreated to record lows (a 30 percent loss over the previous 25 years) while temperatures above the North Pole rose 30 to 50°F above normal. Speaking to the *Guardian*, atmospheric scientist Jennifer Francis warned: "This is all headed in the same direction and picking up speed."

On December 1, 2016, top military officials from the United States, the United Kingdom, and Bangladesh issued a "call to arms" on climate change during a Chatham House conference hosted by London's Royal Institute of International Affairs. Brigadier-General Stephen Cheney, CEO of the American Security Project, warned that "climate change could lead to a humanitarian crisis of epic proportions," triggering crop failures, water scarcity, and devastating storms, while British rear admiral Neil Morisetti called climate change "a strategic security threat that sits alongside others like terrorism and state-on-state conflict."

When the U.S. electoral college selected Donald Trump to serve in the Oval Office, many senior military and national security officials started closing ranks to deal with a commander-in-chief who has ignored the evidence for climate change (a "hoax created by and for the Chinese") and has called for creating "many millions of high-paid jobs" by rolling back pollution regulations, attacking the historic 2016 Paris Climate Accord, and reviving the "Smokestack Economy"—with increased reliance on coal, oil, and fracking.

Meanwhile, with entire hemispheres at risk, the hammer-blows of war continue to fall, in large measure, on ecosystems, villages, and cities. This is the initial focus of Part II, *Terracide—The War on Nature.*

Mapping the Terrain

Nature in the Crosshairs offers a short survey of "scorched earth" warfare and visits conflict zones on several continents. These essays—many written by activists who were eyewitnesses to the destruction they describe—range from Afghanistan to Africa, from Central America to the Middle East, from distant Pacific islands to the chemically poisoned valleys of Vietnam.

Collateral Damage examines war's impacts on the "urban battlescape"—in Baghdad, Ukraine, and Syria—and considers the nameless victims of conflict, from refugees and survivors to the ocean's largest marine mammals.

A Field Guide to Militarism chronicles how war has metastasized beyond the traditional battlefield, expanding to threaten native lands, deserts, oceans, islands, the Arctic, and even the vastness of outer space.

The Machinery of Mayhem exposes the environmental impacts of military operations and weaponry, including tanks, warships, jet aircraft, cluster bombs, landmines, weather modification, and nuclear war.

NATURE IN THE CROSSHAIRS

How can you conserve nature when you are bombing nature in wars of choice around the world, practicing military operations in areas that have endangered species like on the islands of Oahu, Big Island of Hawaii, Pagan, Tinian, and Okinawa, and bombing islands into wastelands like the Hawaiian island of Koho'olawe and the Puerto Rican island of Vieques and now you want to use the North Marianas "Pagan" Island as a bombing target?

—Col. Ann Wright (U.S. Army, Ret.)

Afghanistan: Bombing the Land of the Snow Leopard

Joshua Frank (2010)

Shipping off 30,000 troops to the land of the Taliban may be infuriating to devoted antiwar activists, but the toll the Afghanistan war is having on the environment should also force nature lovers into the streets in protest.

Natural habitat in Afghanistan has endured decades of struggle, and the "war on terror" has only escalated the destruction. The lands most afflicted by warfare are home to critters that most Westerners only have a chance to observe behind cages in our city zoos: gazelles, cheetahs, hyenas, Turanian tigers, and snow leopards, among others. Afghanistan's National Environmental Protection Agency, which was formed in 2005 to address environmental issues, has listed a total of thirty-three species on its endangered list.

In 2003, the United Nations Environment Programme (UNEP) released its evaluation of Afghanistan's environmental issues. Titled *Post-Conflict Environmental Assessment*, the UNEP report claimed that war and longstanding drought "have caused serious and widespread land and resource degradation, including lowered water tables, desiccation of wetlands, deforestation and widespread loss of vegetative cover, erosion, and loss of wildlife populations."

Ammunition dumps, cluster bombs, B-52 bombers, and land mines (which President Obama refused to ban) serve as the greatest threat to the country's rugged natural landscape and the biodiversity it cradles. The increasing number of Afghanis that are being displaced because of military conflict, UNEP's report warned, has compounded all of these problems. It was a sobering estimation. However, it was an analysis that should not come as much of a surprise: warfare kills not only humans, but life in general.

As bombs fall, civilians are not the only ones put at risk, and the lasting environmental impacts of the war may not be known for years, perhaps decades, to come. For example, birds are killed and sent off their migratory course. Literally tens of thousands of birds leave Siberia and Central Asia to find their winter homes to the south. Many of these winged creatures have traditionally flown through Afghanistan to the southeastern wetlands of Kazakhstan, but their numbers have drastically declined in recent years.

Endangered Siberian cranes and two protected species of pelicans are the most at risk, say Pakistani ornithologists. Back in 2001, Dr. Oumed Haneed, who monitors bird migration in Pakistan, told the British Broadcasting Corporation that the country had typically witnessed thousands of ducks and other wildfowl migrating through Afghanistan to Pakistan. Yet, once the United States began its bombing campaigns, few birds were to be found.

"One impact may be directly the killing of birds through bombing, poisoning of the wetlands or the sites which these birds are using," said Haneed, who works for Pakistan's National Council for Conservation of Wildlife. "Another impact may be these birds are derouted, because their migration is very precise. They migrate in a corridor and, if they are disturbed through bombing, they might change their route."

Targeting the White Mountains

Intense fighting has raged throughout Afghanistan. The White Mountains, where the U.S. hunted Osama bin Laden, has been hit particularly hard. While the difficult-to-access ranges may serve as safe havens for alleged al-Qaeda operatives, the Tora Bora caves and steep topography also provide refuge for bears, Marco Polo sheep, gazelles, and mountain leopards. Every missile that is fired into these vulnerable mountains could potentially kill any of these treasured animals, all of which are on the verge of becoming extinct.

"The same terrain that allows fighters to strike and disappear back into the hills has also, historically, enabled wildlife to survive," Peter Zahler of the Wildlife Conservation Society told *New Scientist* at the onset of the 2001 Afghanistan invasion. Zahler also warned that not only are these animals at risk from bombing, they are also at risk of being killed by refugees. The fur from an endangered snow leopard (whose population is said to number between 100 and 160) can score $2,000 on the black market. That money, in turn, can help displaced Afghanis pay for safe passage into Pakistan.

As Zahler observes, "The story in Afghanistan is not the actual fighting. It's the side effects—habitat destruction, uncontrolled poaching, that sort of thing." Afghanistan has faced nearly thirty years of unfettered resource exploitation, even prior to the start of the 2001 war. This has led to a collapse of government systems and has displaced millions of people, all of which has led to the degradation of the country's habitat on a vast scale. Forests have been ravaged to provide short-term energy and building supplies for refugees. Many of the country's arid grasslands have also been overgrazed and wildlife killed.

"Eventually the land will be unfit for even the most basic form of agriculture," explained Hammad Naqi of the World Wide Fund for Nature in Pakistan. "Refugees—around four million at the last count [in 2001]—are also cutting into forests for firewood."

The Return of Carpet-Bombing

In early 2001, during the initial attacks, the BBC reported that the United States had been carpet-bombing Afghanistan in numerous locations. John Stufflebeem, deputy director of operations for the U.S. Joint Chiefs of Staff, told reporters at the time that B-52 aircraft were carpet-bombing targets "all over the country, including Taliban forces in the north. We do use [carpet-bombing strategies]. We have used it and will use it when we need to." (The George W. Bush White House has denied that it implemented this destructive strategy.)

Additionally, Pakistani military experts and others have made allegations that the United States has used depleted uranium (DU) shells to strike targets inside Afghanistan, most notably against the Taliban front lines in the northern region of the country. Once these deadly bombs strike, they rip through their target and erupt into a toxic cloud of fire. Many medical studies have shown that DU's radioactive vapors are linked to leukemia, blood cancer, lung cancer, and birth defects.

"As U.S. and NATO forces continue pounding Afghanistan with cruise missiles and smart bombs, people acquainted with the aftermaths of two recent previous wars fought by the U.S. fear—following the Gulf and Balkan War Syndromes—the Afghan War Syndrome," Dr. Ali Ahmed Rind has observed. "This condition is marked by a state of vague aliments and carcinomas, and is linked with the usage of Depleted Uranium as part of missiles, projectiles, and bombs in the battlefield." France, Italy, and Portugal asked NATO to halt DU use, but the Pentagon still does not admit that DU is harmful or that it has used such bombs during its assaults in the country.

Afghanistan's massive refugee crisis, lack of governmental stability, and extreme poverty, coupled with polluted water supplies, drought, land mines, and excessive bombings, all contribute to the country's intense environmental predicament. Experts unanimously agree: there simply is no such thing as environmentally friendly warfare.

Africa: Wars on Wildlife

Jane Goodall, PhD, DBE (2003)

In 1960, I began my study of chimpanzees on the shores of Lake Tanganyika in what is now Tanzania. At that time, chimpanzee habitat stretched for miles, fringing the lake from Burundi to Zambia in the south. From the hills of Gombe Stream National Park, one could see the forest stretching away inland, interrupted only by a few villages with their fields of crops.

Today, the scene is very different: cultivated land crowds up to the boundaries of the park, the trees have gone, peasants are trying to grow crops on the steep rocky hillsides, causing terrible erosion, the soil is losing its fertility, the forest animals have gone, and the human population is struggling to survive.

What has caused this devastation? Partly, of course, the same kind of population growth that we have seen around the world since 1960. But the situation has been made infinitely worse by the vast numbers of refugees fleeing the wars ravaging Burundi, to the north, and the Democratic Republic of the Congo (DRC) on the other side of the lake, to the east.

Refugees in Africa, as they trudge toward some place of safety—usually one of the United Nations High Commissioner for Refugees camps—have a terrible struggle to survive. They cut down trees for temporary shelter and for firewood, gather every kind of edible plant, and hunt wildlife for food. This fuels tensions between the local people and the refugees. Scarcity of natural resources can actually trigger conflicts as well as prolonging existing wars.

Wild animals (as well as livestock) are often direct casualties of war. Soldiers as well as refugees hunt wildlife for food. According to the Biodiversity Support Program, war in the DRC in 1996 and 1997 led to an escalation in poaching in one area that reduced the elephant population by half, buffalo by two-thirds, and hippo by three-quarters. Many other animal species were also affected. And this included the great apes—gorillas, chimpanzees, and bonobos—already seriously endangered by the commercial bush meat trade.

Wildlife, as well as innocent humans, are often maimed by land mines. Hundreds of animals are also affected, and vast areas of farmland are made useless so that increased destruction of wilderness areas by desperate people results.

The instability caused by conflict enables people to take advantage of the situation to mine diamonds and other commercially valuable resources illegally, even in protected areas—where they destroy the environment and kill all available wildlife for food.

Elephants are one of the most amazing animals on earth, yet unless we get together to help protect them, they could become extinct. Because of us.

African elephants, today, are victims of the ivory trade. They are being slaughtered by the thousands—on average one hundred each day—many killed by commercial poachers who fuel the ivory trade, others by armed militias that often use poached ivory to fund their wars. Save the Elephants (the NGO founded by Iain Douglas-Hamilton) estimates an average of 33,000 elephants were lost to poachers every year between 2010 and 2012. Over 64 percent of elephants in Central Africa have vanished in the past decade. And in Tanzania, where poaching has been shocking, the estimated number of individuals has dropped from almost 110,000 to just over 43,000 in the past five years.

How cruel we are as a species. Each of the thousands of elephants murdered had his or her own personality, place in the herd, family bonds. Each was an individual that mattered. Each deserved respect. Each was destroyed to hack out ivory—even from juveniles with tiny tusks. Bodies butchered, family bonds broken, and the desperate grief and trauma of those left behind—those spared because they had no tusks, or because they managed to escape.

The fight can't be on just one front. It starts on the ground, with the courage of the rangers fighting the poachers, but it extends to the halls of power through corrupt traders, businessmen, police, and even politicians. Experts estimate that elephants will be virtually extinct within the foreseeable future if we don't take action. The United States and China are the two biggest markets and, given its role on the world stage it is encouraging that the Chinese government has agreed to ban all trade in ivory, both internationally and within the country, by the end of 2017. Let us hope that other countries will follow suit. And that all trophy hunting will also be prohibited. [In June 2016, the Obama administration, acting under the Endangered Species Act, announced a near-total ban on the interstate trade in ivory.]

El Salvador: Scorched Earth in Central America

Gar Smith (1988)

Helicopters and screeching jets descend over a village. Bombs explode, carving craters. Families crying out in terror scramble to find shelter. Children caught in billowing walls of fire run screaming down dirt roads, clothing burned from their bodies. In less than an hour, a village has been destroyed, 100 civilians have been killed, more than 4,000 are homeless.

This was not Vietnam. This was El Salvador.

On the tenth anniversary of America's retreat from Vietnam, the weapons and strategies of America's most unpopular war were reappearing—with a vengeance—in El Salvador. Huey helicopters, A-1 howitzers, blockbuster bombs, fragmentation weapons, search-and-destroy missions, "free-fire zones," white phosphorous, and even napalm were bringing death and disorder to the people and environment of Central America.

By 1988, one-tenth of El Salvador's population had fled the country, according to Americas Watch. More than a million people—one-fifth of the populace—had become refugees. According to a report from the Committee in Solidarity with the People of El Salvador (CISPES): "People leave what little they own, to run by night and hide by day, living on roots and stifling their children's cries so as not to be detected by the troops." Still, the refugees may be the lucky ones. For those left behind in provinces re-designated "free-fire zones," survival cannot be guaranteed.

In the Invisible War, people suffer. So does the land. The Pentagon's "secret" air war obliterated forests as well as villages while the ground war slaughtered wildlife and livestock, disrupting environmental stability.

Bombs Bursting in Air

The major weapons used in the air war against the Farabundo Marti National Liberation Front (FMLN) were the 250-pound, 500-pound, and 750-pound bomb. Unfortunately, these bombs don't discriminate between guerilla fighters and peasant farmers. In 1988, CISPES reported that "almost 2,000 civilians have died in these air raids since January 1984."

Since the bombing began in October 1982, the air war became, in the words of CISPES, "the largest escalation of U.S. military involvement in foreign lands since the Vietnam War." It was clearly the largest air war in the history of Latin America. In 1982, human rights groups recorded 111 confirmed bombings of civilian targets in El Salvador: in 1984, that number more than tripled with reports of 338 attacks.

In September 1984, a delegation of American doctors and health professionals traveled behind army lines into the Salvadoran countryside. The delegation from Medical Aid to El Salvador (MAES) visited one of the "military targets" but discovered "no recent military activity was evident . . . the bombs were very direct hits on what were clearly civilian homes, producing two dead children and several wounded civilians; neighboring civilian homes were all flying white flags."

Villagers interviewed in Miramundo, in Chalatenango Province, said they had been harassed by bombing raids for four years.

As Father Jesus Nieto, a Salvadoran priest, observed: "No shelter can protect the people from a 500-pound bomb. These leave craters fifteen-feet deep and sever trees too thick to encircle with one's arms. . . . Imagine what such bombs do to frail human bodies. They are blown way without a trace."

The effect on peasant farming was devastating. "We are facing a food crisis," Father Nieto warned, "for when the bombs fall on the fields, the crops are totally destroyed."

In 1987, SALPRESS (an independent press organization in El Salvador) reported: "Between April 6 and 12 in the province of Cuscutlan, Air Force bombs destroyed 286 houses, killed more than 116 head of cattle, and burned the entire corn and grain crops and most of the zone's fruit trees."

"Iron Bombs"

When an "iron bomb" crashes into the ground, 30 percent of the explosive force is released into the soil. To get more "blood per bang," the Pentagon attaches three-foot-long "fuse extenders" to trigger the explosions slightly *above* ground level to increase the killing effect.

In Vietnam, these "antipersonnel" bombs were called "Daisy Cutters." Whatever the name, the result is the same: they cut off everything at ground level—trees, crops, livestock, people. While regular demolition bombs were used to destroy buildings, iron bombs were dropped in open spaces.

According to the Lawyers Committee for International Human Rights and Americas Watch, "the attacks on civilian non-combatants . . . are part of a deliberate policy . . . to force civilians to flee, depriving the guerillas of a civilian population from which they can obtain food and other necessities." Even during the supposedly "quiet" period during the March 1984 Salvadoran election, CISPES claimed the Air Force dropped "7.5 tons of explosives a day on rural villages." The group claimed that 235 civilians were killed in 3 small villages in a single 13-day period.

The Return of Napalm

In towns throughout the countryside, bombing survivors interviewed by MAES and others told stories of a bomb that erupts in boiling jets of flame and collapses into clouds of dark smoke that sucks the oxygen from the air and stains the earth with a dark, oily film. MAES doctors who examined wounded survivors had no doubt what had been dropped. They'd seen the effects before, in Vietnam.

It was napalm.

"We do not have incendiary bombs," claimed Col. Ricardo Aristdes Cienfuego, speaking on behalf of the Salvadoran Joint Chiefs of Staff. But near Tenancingo, a solider with the Atlacatl Battalion informed a Pacific News Service reporter that incendiary bombs are routinely used in advance of major ground operations.

"Usually we drop incendiary bombs before we begin operations in the area," stated another soldier with the 5th Infantry Brigade. "By the time we enter the area, the land has been burned over and the subversives pretty well toasted."

The reporter managed to interview one of the "subversives" who told a different story. "I was outside my house when the bomb fell," the woman stated. "I could not see anything because of the black smoke and could not get air. Everything was on fire. My two children burned to death."

In a May 26, 1983, congressional report, Col. Rafael Bustillo, the commander of El Salvador's Air Force, confessed: "We had to use napalm because we didn't have any other equipment. We bought if from Israel several years ago and used it until 1981."

"Napalm leaves behind a gooey residue on the earth," explains MAES national coordinator Christina Courtright. "This residue apparently doesn't disintegrate." She mentioned a report by a Mexican physician who returned to a napalmed site three years after an attack and found "nothing had grown on this patch of ground, years after the event."

Elemental Fire

"We do not supply incendiary bombs to the Salvadorans," U.S. Embassy spokesperson Greg Lagana told the *Christian Science Monitor* on April 27, 1984. "The only incendiary device is the white phosphorous rocket used to mark an area for bombing."

White phosphorous is a soft, pliable, gray metal that is so highly reactive it burns spontaneously when it comes in contact with air. In water, it explodes. White phosphorous is used to fill 2.75-inch rockets fired to mark targets with smoke and a white flame. Sometimes the targets are human.

One young woman from Morazan graphically described a phosphorous injury: "Bits of burning material from an explosion bored deep into the victim's arm, producing light-colored smoke and little flame. Mud was applied to smother the flame but it quickly became dried out by the heat produced inside the wound and the mud had to be replaced continuously. The burning stopped one day later, after repeated applications of mud." People treating the wounded have to cover their faces with damp cloths to fight the toxic fumes.

"When burning, white phosphorous enters the body, and it keeps on burning," according to Harvard University biochemistry professor Dr.

Matthew Meselson. "It will burn under water and actually burn inside the body. It can be a horrible antipersonnel weapon." Even after it has burnt itself out, phosphorous remains deadly. Inhaled or taken into the bloodstream, it can prove lethal in as little as two days.

Further testimony that white phosphorous had been used came from former U.S. Air Force pilot, Vietnam veteran, and physician Dr. Charles Clements, who visited the Guazapa area in 1982–83. Clements claims that the "military targets" in that area included elementary schools and medical clinics.

How War Is Killing the Land

El Salvador, which has lost much of its rich and diversified forests to agriculture and cattle-ranching, now faced a new threat from the air and ground war. Native plant species like the *chicozapote, sincuya,* and *cabeza de muerto* were placed on the path to extinction. Rubber trees and willows also were threatened. Native wildlife, threatened by the pressures of troop movements and bombardment, included forest-dwelling sloths, tiger cats, howling monkeys, iguanas, and deer.

El Salvador is a part of the great Central American refuge that serves as a wintering home for millions of migrating North American songbirds—tanangers, warblers, kingbirds, and vireos. More than 150 species of North American birds are dependent on a rainforest refuge that lost 75 percent of its cover in the three decades prior to the 1960s. Scientists have noticed fewer songbirds returning in the spring migrations.

"Natural resources are a specific target of military action," writes Mexican researcher Maria del Carmen Rojas Canales in her study, "Ecological Effects of the War in El Salvador." According to Rojas Canales, "one of the departments most affected is that of Cuscatlan, both because of the type of arms used and the intensity of the attacks." Cuscatlan is largely semi-evergreen rainforest with croplands devoted to corn, beans, sugar cane, zapote, mamey, mango, nispero, nance, and marañón. "The types of attacks are bombing with napalm and white phosphorus, 200-pound bombs, burning of cane plantations, and destruction of coffee crops," Rojas Canales observes. Fruit trees damaged by bomb fragments become susceptible to diseases caused by bacteria and fungi. The effect of incendiary weapons also works to "diminish plant cover and lead to soil erosions, which would be accelerated due to the heavy rainfall of the area." Beyond the problem of erosion, "the entire ecological balance is damaged by extinguishing the primary producers that serve as food and refuge for many other organisms such as rodents, many birds, and other creatures which, if not of commercial interest, do contribute to the maintenance of the balance of the ecosystem."

Furthermore, war-caused fires in the pine-oak forests "provoke changes in the composition and the structure of the ecological communities," causing the forests to decline into secondary "scrub forests" or grasslands, which leads to more erosion, "the drying up of natural springs, water pollution, floods, filing up of reservoirs, and dust storms."

"If the war continues and intensifies the use of conventional as well as chemical weapons," Rojas Canales concludes, "the deterioration could reach a point that would impede the recuperation of ecosystems even in the medium term. There would be zones that, as in the case of Vietnam, would be unusable for more than fifty years."

Bombs, Bullets, and Bats

In March 1985, Gus Newport, the mayor of Berkeley, California, attempted to visit San Antonio Los Ranchos, Berkeley's sister city in Chalatenango Province. The city had been nearly destroyed by a bombing raid in the fall of 1983. Newport and a colleague, Diane Green, never reached their goal. In the middle of their cross-country hike, their party came under a government air attack.

Newport and Green huddled in a dark bunker with a dozen terrified children as the rockets pounded the village they had taken shelter in. One rocket destroyed a nearby home, killing the family of eleven cowering inside.

"The army is destroying fields and homes. They have destroyed anything that is living," Newport told a press conference upon his return. Fighting back tears, Newport confided his fears that he "might never see again" many of the friends he had made during his visit to Chalatenango.

Diane Green brought back vivid memories of the environmental havoc raging over the countryside: "We observed lots of land that had been burned by mortars. They can start fires that can burn for days. The army drops incendiary rockets, gasoline rockets. We saw lots of land—acres and acres—that had been charred."

"When the Salvadoran army invades," she stated, "they always burn the crops and destroy the water pipes in the villages. They cut down the fruit trees, kill the cows, destroy all the tools they find. Fishing boats are sunk." Both Newport and Green reported seeing many adults and children suffering from malnutrition as well as machine-gun wounds.

"This is what counterinsurgency is all about," says Christina Courtright. "The government has tried very hard to destroy the means of subsistence of the people in the countryside. They have intentionally destroyed crops and homes. Dead bodies are thrown into wells to poison the water supplies.

"Ecological imbalances are occurring all over," Courtright continued. "One of the most macabre involves bats." In Central America, it is a common sight to see blood trickling down the necks of cattle every morning. It is part of the natural ecology. The war has changed that, Courtright says. "Vampire bats are now attacking humans for the first time because the army has killed off the cows.

"The bats first lick the victims' necks in the night," Courtright explains. "Their tongues have a mild anesthetic so the victim doesn't feet the bite. But the bite also injects an anticoagulant. This is fine for the bat but it's not good for the victim if the victim happens to be a small child. Small children can die from loss of blood in the night. And they have."

Guam: The Tip of America's Global Spear

Catherine Lutz (2010)

The island of Guam is a most remarkable place—a land of cultural distinctiveness, resourcefulness, and great physical beauty. But the Chamorro people who have lived here for 4,000 years have a historical experience with colonialism and military occupation more long-lived and geographically intensive—acre for acre—than anywhere else in the Pacific. Today

Guam is embroiled in a debate over when, how, or if, the United States military will acquire additional land to make more intensive use of the island.

This military expansion was planned in Washington, with acquiescence and funding from Tokyo, in order to relocate some 5,000 Marines and their dependents (along with U.S. Navy, Army, and Air Force assets and operations) from the Japanese island of Okinawa to Guam and the Commonwealth of the Northern Marianas. The original expansion plans for Guam included: removal of 71 acres of coral reef from Apra Harbor to allow the entry and berthing of nuclear aircraft carriers; the acquisition of land (including an ancient and revered Chamorro village at Pagat) for a live-fire training range; and an estimated 47 percent increase in the island's population—already past its water-supply carrying capacity. (These figures have changed in the last few years and are still in flux. After vigorous local pushback, some DC politics, and a court ruling, the amount of land involved has been reduced.)

The Pentagon has offered to scale back its expansion plans. Instead of cutting down 400 acres of forest to build homes for 1,500 military dependents, the families will now be housed at Andersen Air Force Base. The military expansion is being undertaken with nearly one-third of the island already in military hands and a substantial historical legacy of environmental contamination, resource depletion, external political control, and other problems brought about by the existing military presence.

Pushback has been substantial, something that is particularly remarkable given that many islanders consider themselves loyal and patriotic Americans and many receive military paychecks or pensions as soldiers, veterans, or contract workers. Dissent among Guam's social sectors rose dramatically in November 2009 after the release of a draft Environmental Impact Statement (EIS) revealed how extensive Washington's plans for the island were. It became clear that Guam's political leaders and citizens were to be simply *informed* of those plans, rather than consulted—let alone asked for permission to use the various assets. The Environmental Protection Agency subsequently found the EIS to be a deeply inadequate document that failed to present a fair and clear assessment of the environmental costs of the military's plans. A demonstration at a sacred site at Pagat on July 23, 2010, provided a potent symbolic expression of the growing popular resistance to the base plan.

The people of Guam are waiting to hear exactly how many more acres of their land will be taken for military purposes, how many tens of thousands of new people and new vehicles will be visited on the island, how many military overflights and aircraft carrier visits will follow, and how many toxic trickles or spills will be imposed upon them.

This is Guam's colonial history and colonial situation. It is colonial even as many of Guam's residents take their U.S. citizenship seriously. The U.S. presence in Guam is properly called "imperial" because the United States is an empire in the strict sense of the term, as used by historians and other social analysts of political forms.

Besides colonialism, another concept relevant to Guam's situation is militarization. It refers to an increase in labor and resources allocated to military purposes and the shaping of other institutions in synchrony with military goals. It involves a shift in societal beliefs and values in ways that legitimate the use of force. It helps describe the process by which fourteen-year-olds are placed in uniforms and required to carry proxy rifles in JROTC units in all of Guam's schools. It explains why a fifth to a quarter of high school graduates enter the military and why the identity of the island has, over time, shifted from a land of farmers to a land of war survivors to a land of loyal Americans to a land that is, proudly, "The Tip of the Spear"—that is, an island that has become a weapon. This historical change—the process of militarization or military colonization—has been visible to some, but more often, it lies hidden in plain sight.

Military base communities are, in many ways, as distinctive sociologically and anthropologically as the military bases they sit next to, because they respond in almost every way to the presence of those bases. They are not simply independent neighbors: over time, they become conjoined, although one is always much more powerful than the other.

These military facilities include sprawling Army bases with airfields, artillery testing ranges, berthed aircraft carriers, small listening posts, schools, golf courses, basketball courts, and McDonald's. While the bases provide troop barracks, weapons depots, and staging areas for war-making and ship repair facilities, they are also political claims, spoils of war, arms sales showrooms, toxic industrial sites, laboratories for cultural (mis)communication, and collections of customers who patronize local shops, services, bars, and brothels.

The environmental, political, and economic impact of these bases is enormous. While some people benefit from the coming of a base (at least temporarily), most communities pay a high price: their farmland taken for bases; their bodies attacked by cancers and neurological disorders from exposure to military toxins; their neighbors may even be imprisoned, tortured, or disappeared by the autocratic regimes that survive on U.S. military and political support given as a form of tacit rent for the bases.

The count of U.S. military bases should also include the eleven aircraft carriers in the U.S. Navy's fleet, each of which is referred to as "four-and-a-half acres of sovereign U.S. territory." These moveable bases and their land-based counterparts are just the most visible part of the larger picture of U.S. military presence overseas. This picture includes (1) U.S. military training of foreign forces, often in conjunction with the provision of U.S. weaponry, (2) joint exercises meant to enhance U.S. soldiers' exposure to a variety of operating environments from jungle to desert to urban terrain and interoperability across national militaries, and (3) legal arrangements made to gain overflight rights and other forms of ad hoc use of others' territory as well as to preposition military equipment. In all of these realms, the United States is in a class by itself with no adversary or ally maintaining anything comparable in terms of its scope, depth, and global reach.

Foreign military bases have been established throughout the history of expanding states and warfare. They proliferate where a state has imperial ambitions, either through direct control of territory or through indirect control over the political economy, laws, and foreign policy of other places. Whether or not it recognizes itself as such, a country can be called an empire when it projects substantial power with the aim of asserting and maintaining dominance over other regions. Those policies succeed when wealth is extracted from peripheral areas and redistributed to the imperial center. Empires, then, have historically been associated with a growing gap between the wealth and welfare of the powerful center and the regions it dominates. Alongside and supporting these goals has often been elevated self-regard in the imperial power or a sense of racial, cultural, or social superiority.

Alongside their military and economic functions, bases have symbolic and psychological dimensions. They are highly visible expressions of a nation's will to status and power. Strategic elites have built bases as a visible sign of the nation's standing, much as they have constructed monuments and battleships.

Is Guam's Andersen AFB a domestic base or a foreign base? As Guam is a U.S. territory, it is neither a fully incorporated part of the United States nor a free nation. The island's license plate, which notes it is "Where America's Day Begins," also reads, "Guam USA." This expresses the wish of some, rather than the reality. It perhaps would better read, "Guam, U.S. sort of A." International legal norms make the status clear, however. Guam is a colony, and primarily a military colony, in keeping with the idea that the United States' imperial history, especially in the second half of the twentieth century, has been one of military colonialism around the world.

Hawaii is considered a heavily militarized state but, while the Pentagon exercises control over 5.7 percent of Hawaii's 6,423 square miles, it currently commands 29 percent of Guam's 212 square miles. No Status of Forces Agreement (SOFA) regulates the U.S. forces on Guam. The DoD does not need to report each day to the government of Guam on how many soldiers have been brought in or sent out of Guam, nor is it negotiating with Guam about its plans to grow its bases on Guam.

One very important and empirical index of the degree to which Guam's bases are foreign or domestic is the quality of care that has been taken with its environment and health. Overseas bases have repeatedly inflicted environmental devastation. Unexploded ordnance killed twenty-one people in Panama before the United States was evicted and continues to threaten communities nearby. In Germany, industrial solvents, firefighting chemicals, and varieties of waste have ruined ecological systems near some U.S. bases. The Koreans are finding extremely high levels of military toxins in bases returned to them by the United States.

Activists have long considered the environmental and judicial standards that are negotiated into each country's SOFA as an index of how much respect their country is accorded. It is possible to measure the quantity of toxins variously introduced into the environment of Guam, Germany, the Philippines, California, and North Carolina, for example. The broad differences in that quantity roughly occur on a scale that appears quite racial, with the U.S. mainland at the top, Germany next, and the Philippines and Guam at the bottom. If Guam's political status were truly domestic, we might expect Guam to look more like the mainland in terms of how the environment has been cared for. It does not.

But the internal racial history of the United States itself demonstrates that the military base has been a booby prize for many of the internally colonized in the United States as well. All domestic military bases are, in fact, built on Native American land and even after that land was taken, the bases were often intentionally sited on land inhabited by poor white, Black, and Indian farmers. Thousands of them lost their land in North Carolina alone in the buildup to World War II.

The military publicizes its arrival or expansion as an economic boon, noting the dollars brought in via soldier's salaries, civilian work, construction, and other subcontracts that provide jobs. The First Hawaiian Bank once published a Guam Economic Forecast that claimed: "The military expansion is anticipated to benefit Guam's economy in the amount of $1.5 billion per year once the process begins."

The reason for the economic focus on base expansion is that these impacts appear overall to be positive—unlike the environmental, sovereignty, cultural, crime, and noise effects. One of the reasons these look positive is because the powerful people who benefit from development have the resources to convince others that they, too, will benefit—even when they palpably do not. Typically, however, retail jobs are the main type of work created around military bases. Because these jobs pay less than any other category of work, they only serve to accelerate the growth of inequality in military communities.

Meanwhile, the military is a highly toxic industrial operation that *externalizes* many of its costs of operation to the communities that host it and serve it. These costs include environmental pollution, PTSD in returning war veterans, high rates of domestic violence, rapid deterioration of roads and other public amenities, and, in many communities, a decline in human capital development of populations that have gone into the military.

Finally, military economies are volatile. While the "war cycle" is different than the business cycle, it also has booms and busts. Any major deployment from Guam's bases can be expected to significantly harm local enterprise dependent on military business. Moreover, a volatile real estate market catering to foreign military personnel sends property prices spiraling and forces local working families into substandard housing.

There are legal questions in the Guam military buildup as well. In her testimony before the UN Special Committee of the 24 on Decolonization in 2008, Sabina Flores Peres referred to the extremity of "the level and grossness of the infraction" of the UN Charter by the United States in its further militarization of the island. This is not hyperbole. Guam's militarization is objectively more extreme in its concentration than that found virtually anywhere else on earth. There are only a few other areas that are in similar condition—all, not coincidentally, islands such as Okinawa and Diego Garcia.

Guam, objectively, has the highest ratio of U.S. military spending and military hardware—and land-takings from indigenous populations—of any place on Earth. The level of the injustice meted out to island residents has to do with the racial hierarchy that fundamentally guides the United States in its "negotiations" with other peoples over the siting of its military bases and the treatment they are accorded once the United States settles in.

Kuwait: The War That Wounded the World

William Thomas, Gulf Environmental Emergency Response Team (1991)

In March 1991, shortly after the world's biggest oil spill began washing ashore at Al-Jubail, I was in Dammam at the Saudi environmental headquarters watching the first hard numbers listing all known sources of oil washing into the Persian Gulf scroll across an American scientist's computer screen. The 5.5 million barrels total—equal to 25 *Exxon Valdez* spills—far eclipsed the largest previous oil spill, which released 3.3 million barrels off the Mexican coast in 1979. "One-third of this oil," the American told me, "is the result of allied bombing."

Flying through the eerie midday darkness, I made two coastal survey flights with the Royal Saudi Air Force. Shivering in the unusual bitter cold, the stench of oil clogging our nostrils, we flew hour after hour at 100 knots. Not for a single moment did the American and Saudi observers fail to see heavy crude washing ashore from past slicks extending beyond the horizon. Half of the Saudi coastline was heavily oiled; half of the precious mangroves and salt marsh nurseries were destroyed.

When the helicopter landed, no one could speak. Traversing the heart of this blazing darkness, we could clearly see that a few migrating birds had survived this tarry gauntlet. Walking almost daily through heavily mined coastal areas, our three-man survey party would count—over the next five weeks—perhaps several thousand migrating birds. None could be saved.

We could have used some help ourselves. In the weeks and months following the "liberation" of Kuwait, four members of the Gulf Environmental Emergency Response Team, the U.S. conservation organization Earthtrust, and the World Federation for the Protection of Animals were the sole international response to the biggest environmental catastrophe in modern times.

While one volunteer set about saving the starving survivors of Kuwait's zoo and four fear-crazed horses from the emir's Summer Palace in the center of the Great Burgan oil field, Earthtrust team leader Michael Bailey, ornithologist Rick Thorpe, and I opened the Kuwait Environmental Information Center in the lobby of the city's only functioning hotel.

Although a United Nations Environment Programme press briefing insisted that the choking midday darkness posed "no danger to human health," official U.S. and Kuwaiti assurances were hard to accept after flocks of birds began dropping dead onto city streets. When autopsied sheep in the town of Ahmadi revealed lungs as hard and black as shoe leather, people began asking: "What is happening to *us*?" Though no test findings were being released by Western monitoring teams, Boston's National Toxics Campaign reported unsafe levels of 1.4-dichlorobenzene, arsenic, zinc, cadmium, and lead 175 miles downwind from Ahmadi.

Six weeks after victory was declared, admissions soared at the Ahmadi hospital. Almost all Ahmadi residents claimed to be suffering adverse health effects from more than 400 oil wells burning nearby. While flames from burning oil wells across the street sent black smoke into the petroleum pall, a senior administrator interrupted his denials of danger with frequent bouts of coughing.

The Oil Comes Ashore

On April 26, 1991, I was aboard a survey flight that spotted a massive new slick moving south from Iraq. Loading scavenged oil booms onto a truck provided by the U.S. Army's 352nd Division, Rick Thorpe and I and an Egyptian chef from the hotel spent five days dragging the heavy boom across the fast-flowing estuary at al-Khiran. We were in time to save Kuwait's main wetlands from the brunt of the slick that soaked most of the coastline. By then, more than 460 miles of Arabian Gulf coastline had been heavily oiled. Despite the efforts of U.S. Marines who cleaned the beaches of Kara Island as the first green and hawksbill turtles begin coming ashore to bury their eggs, hundreds of endangered turtles had been found dead in oily inshore waters.

In addition to the toxicity of oil and its smothering effects on mangrove and seagrass breeding grounds, the decrease of 10 to 18°F over smoke-shaded Gulf waters further affected marine reproduction, leading to a decline in female turtle hatchings.

Even before the war, the Arabian Gulf was the world's most polluted waterway. With 25 large oil terminals loading more than 20,000 tankers a year, oil spills of 150,000 metric tons were regarded as routine. Oil has been used as a weapon in the Middle East since biblical times. The disastrous Iran-Iraq War, which saw the rocketing of Iran's Nowruz drilling platform in 1983, released more than a half-million barrels of oil into a shallow sea.

By the time Iraqi forces began pumping oil from Kuwait's Sea Island loading terminal directly into the sea, and allied warplanes had started bombing oil tankers and shore installations in the northern Gulf, Nowruz oil had congealed to the consistency of hard asphalt on the beaches in Qatar and Bahrain.

When thousands of floating mines prevented the deployment of the *al-Wasit* oil recovery ship, Thorpe and I ran a salvaged skiff upcoast into the intake arms of one of Kuwait's three desalination plants. The oil slick we found there was potentially deadly.

Black Rain, "Tarcrete," and Desert Dust

Briefing members of the Kuwait cabinet and royal family, Thorpe, Bailey, and I warned that accelerating desertification and spreading oil lakes, combined with prolonged lack of sunlight, temperature drops of up to 23°C

[73.4°F], and sticky black rain were seriously disrupting the region's ecology. National Oceanographic and Atmospheric Administration (NOAA) chief scientist Dr. Sylvia Earle agreed.

The director of the Center for Remote Sensing at Boston University warned that the disruption of as much as 25 percent of Kuwait's land area by shelling, carpet-bombing, the construction of hundreds of miles of trenches and revetments, and the mechanized movement of more than one million troops would cause redoubling of sand storms. Newly developed fuel-air bombs, whose searing overpressures from exploding gasoline vapor were used to detonate minefields, also pulverized topsoil and completely wiped out vegetation over a wide area. It was no surprise when the shamal winds, which bring the great dust storms every May and June, carried much heavier volumes of sand sweeping over Kuwait immediately after the war. Once anchored by microscopic fungi, Kuwait's disrupted dunes had already begun marching on roads, towns, and cultivated areas following the forty-six-day desert war.

By late 1993, one-third of Kuwait was covered by plant-choking "tarcrete." Oil vapor from the burning wells had mixed with soot and sand and hardened into a four-inch-thick covering. Describing the scale of this latest disaster as "unprecedented in history," U.S. scientist Farouk al-Baz warned that the tarcrete was killing all vegetation, while releasing sand and dust to form more sand dunes. NOAA estimated there were 252 oil pools containing 150 million barrels of oil covering more than half of Kuwait's land surface. Some of these lakes were eight kilometers [five miles] long and eight meters [twenty-six feet] deep. In late April 1991, hundreds of desiccated carcasses were counted around a single oil pool mistaken for water by desperate birds. Within a year, more than one million migratory birds were feared killed after landing on oil-coated wetlands and desert oil pools.

Extensive ordnance pollution continued to haunt both camels and people. Despite extensive postwar mine-clearing operations, Kuwait's shifting desert and beach sands remained infested with Iraqi mines, rockets, and shells. One-third of the 100,000 tons of explosives dropped on Kuwait by coalition forces failed to explode in the soft desert sand. In 1992, as many as 1,600 Kuwaitis were killed or wounded by unexploded ordnance.

Toxic Clouds and a Rain of Bombs

While the oil fires were extinguished by a growing international consortium of firefighters using high-pressure water, chemicals, explosives, and even truck-mounted jet engines, another complication arose with the evaporation of lighter crude "fractions" from each oil gusher. Highly toxic near its source, the invisible vapor dispersed into unknown concentrations over Ahmadi and Kuwait City. There was concern about Kuwait's vast freshwater aquifer, which comes within fifteen feet of the desert surface. The benzene, toluene, and naphthalene found in Kuwait's "sweet" crude saturated those sands with highly toxic carcinogens capable of poisoning the groundwater.

Despite pleas to Gen. Norman Schwarzkopf and other high-ranking U.S. Army commanders, we were unable to get the orders issued that would have mobilized tens of thousands of idle troops and opened up the huge truck parks brimming with bulldozers and other earthmoving equipment awaiting shipment back to United States. Although a European Economic Community (EEC) investigative committee was sympathetic to our call for help, they informed us that revealing the true costs of what had been portrayed by the victorious governments as a "quick and clean" surgical victory was simply too political for EEC involvement.

While Kuwait's dire environmental crisis was largely ignored, the country's suffering could not match the savagery of the air war unleashed against Iraq. The heaviest aerial bombardment ever carried out against another country saw one coalition aerial sortie every 1.8 minutes around the clock. In just six weeks, twice as many high explosives were dropped on Iraq as all the bombs dropped during World War II.

The goading of a madman holding a match over Kuwait's oil wealth risked an ecological holocaust. Despite early warnings—by nuclear winter theorist Dr. Carl Sagan, ozone-hole-discoverer Joe Farman, climatologist Paul Crutzen, and Jordan's top scientific advisor, Dr. Abdullah Toukan—of potential worldwide repercussions if the oil fields were set alight, the leaders of the allied nations decided to roll the dice anyway.

Allied military commanders had already taken a gamble when naval elements of the coalition steamed into the Persian Gulf. In a memo obtained by the Pacific Campaign to Disarm the Seas, senior U.S. Navy officers expressed worry that unprecedented levels of electronic pollution from operation Desert Storm's high-tech offensive could spark a HERO (Hazards of Electromagnetic Radiation to Ordnance) accident, firing the capacitors required to detonate hundreds of nuclear weapons deployed at sea. Even a single nuclear blast could release an electromagnetic pulse capable of triggering a chain reaction among other nukes or conventional warheads primed for attacking the region. The coalition commanders lucked out: no nuclear bombs exploded and the subsequent eruption of the Philippines' Mount Pinatubo volcano on June 15, 1991, conveniently masked the atmospheric effects from the world's biggest oil fire.

"Smart Bombs" and the Highway to Hell

Conventional munitions were devastating enough, but the Gulf War provided an opportunity to field-test some unconventional weapons as well. Advanced weapons such as the thermobaric fuel-air bombs tested in the suburbs and deserts of Iraq were specifically developed by the U.S. weapons industry to mimic the devastation of tactical nuclear warheads. For all their vaunted video footage of laser-guided bombs dropping neatly down air vents and open hangar doors, U.S. military officials later admitted that less than 7 percent of all bombs dropped on Iraq were "smart" or guided bombs. At least 70 percent of the 100,000 tons of bombs dropped on Kuwait and Iraq—smart, dumb, or otherwise—missed their intended targets.

Not shown on officially televised videos were the blackened, limbless corpses of young Iraqi boys who tried to flee Kuwait on the road to Basra. The Canadian Air Force played a leading role in this aerial onslaught. There was nowhere to run when coalition jets began screaming in low over the desert to hit the highway crammed with commandeered taxis, school buses, trucks, and tanks fleeing Kuwait City. Although as many as 15,000 charred corpses, riddled with shrapnel, were hurriedly buried in mass graves by media-conscious coalition forces, the terror of that attack still lingered as I walked the lengthy caravan of burned-out and overturned wreckage lining the Highway to Hell.

As I passed through this junkyard of wrecked vehicles rapidly rusting in the sulfurous air, I kicked at heaps of pitiful plunder: the torso of a doll, costume jewelry spilling from a trunk, lingerie and dresses destined for destitute families and sweethearts back home. I thought of a letter I had found in an Iraqi trench. "Dear brother," wrote one soldier's sister from Baghdad, "We are sorry to hear that you have not been paid in three months and that you have had nothing to eat. Here is some dinar to buy some food."

What about Iraq and the fate of Baghdad, "the world's first city"? Descendants of the Persians and the Samaritans who came before them, the ancient Iraqis gave the West its laws, its alphabet, its calendar, and even the division of time into units of sixty. The Tigris Valley cradled Western civilization.

Soon after the cease-fire was signed, UN observers declared the Tigris River dead. The bombing of Iraq's biggest nerve gas factory, at Samarra, poisoned the main source of the region's drinking and irrigation water. (Sarin, Tabun, and mustard gas are potent chemical warfare agents, capable of saturating concrete and other structures with compounds that can threaten health for years.) Lake Mileh Tharthar—another primary water source—had also been ruined by the destruction of the Samarra plant.

A Lingering, Invisible Shroud of Radioactivity

The UN team also suspected that the Tigris River was radioactive. The U.S. government confirmed the presence of radioactivity in Baghdad following the bombing of a nuclear power plant in a northern suburb of the city. Additional radiation exposure was already haunting Iraqi and Kuwaiti children who played among the forty tons of depleted uranium (DU) rounds left behind on the battlefields by coalition tanks. According to the UK Atomic Energy Authority, there was enough depleted uranium-238 in Kuwait and Iraq to cause "tens of thousands of deaths."

The dense and still-radioactive DU used in antitank rounds ignites on impact and spreads uranium oxide dust that is chemotoxic. DU has been called "the Agent Orange of the '90s." Ingested or inhaled, these radioactive dust particles can trigger kidney disease and cancer. Children are especially vulnerable because their growing cells are rapidly dividing. DNA-altering alpha rays from decaying uranium are also known to pass through the placenta into the developing fetus. In addition to this chemo-radiation danger, cyanide, organochlorines, nitrogen dioxide, dioxins, and heavy metals were released in massive quantities during tens of thousands of bombing raids.

All told, allied pilots leveled four nuclear, twelve biological, and eighteen chemical warfare plants in Iraq. Weather maps obtained by a U.S. Senate aide revealed that much of the toxic debris from those raids was carried downwind over allied troops. The aerial scorched-earth campaign also deliberately targeted Iraqi civilians by hitting at least forty-one power-generating plants and hydroelectric dams—some targets were hit by more than twenty bombs.

Targeting the Support Systems

Croplands, barns, and grain silos were also attacked, along with irrigation floodgates that caused vast incursions of seawater into southern Iraq, killing crops and permanently salting farmland. In the first harvest after the conflict, Iraq suffered almost total crop failure.

The targeting of power plants caused short circuits that burned out Iraq's irrigation pumps. A year after hostilities ceased, disrupted water supplies continued to cause enormous suffering in Iraq, where nearly half of the country's 14 million inhabitants were under the age of 15. A Harvard study team report estimated that waterborne diseases from the deliberate bombing of water and sewage plants could add another 200,000 deaths to the previous tallies of 125,000 dead Iraqi troops, civilians, and Kurdish rebels.

History's most intensive suburban bombing campaign psychologically crippled many young survivors. Nearly two-thirds of the children interviewed by the Harvard study team believed they would never live to become adults. Nearly 80 percent had experienced bombing at close range. One in every four children had lost their homes to allied bombing. Nearly half had lost members of their families in the war.

Because of the intensity of aerial bombardment and economic sanctions following the war, the death rate for children under five tripled. In the first eight months of 1991, 50,000 Iraqi children died. In 1993, the International Commission on the Gulf Crisis (ICGC) estimated that 300,000 children—nearly one-third of Iraq's youth—continued to suffer from malnutrition. As raw sewage submerged the streets of Basra and poured into Baghdad's rivers, roughly 80 percent of water samples taken by the ICGC showed gross fecal contamination. Nearly the entire population was exposed to waterborne diseases. By 1993, typhoid, gastroenteritis, and cholera were epidemic across Iraq. Meningitis, hepatitis, malaria, and polio rates were also up sharply.

A War That Wounded the World

The Iraqis were not alone in their agony. This was "the war that wounded the world." In February 1991—a month before the Kuwait oil wells were ignited by retreating Iraqi forces and before U.S. Marine Harrier "jump jets" dropped napalm on the Greater Burgan oil field—NOAA scientists at Hawaii's Mauna Loa Observatory recorded the sudden surge of atmospheric soot. The fallout was traced back to the first allied air attacks on Basra, seven to ten days earlier.

NOAA's findings were quickly suppressed by the U.S. Department of Energy. The DOE forbade American scientists from releasing their postwar environmental data. The Gulf eco-war cover-up had begun. But within months, independent observers in California, Wyoming, India, China, and Tokyo measured soot particles high in the stratosphere between 10 and 100 times normal levels. Sulfuric oil-soot, characteristic of Kuwait crude, also dropped heavy amounts on Turkey and Oman. Several inches of crude oil covered glaciers high in the Himalayas. Mountain peaks in Germany were also reportedly oiled.

After the EPA insisted that the oil smoke had not risen above 9,000 feet, the *Scientific American* published satellite pictures showing smoke from the fires in Iraq and Kuwait crisscrossing the globe at altitudes up to 33,000 feet. Other satellite images showed a 1,500-mile-long smoke plume extending from Kuwait's blazing oil fields into Iran, Oman, and Pakistan. NASA astronauts reported looking down on a planet enveloped in haze.

One-third of the trees and crops in Iraq's Western province were destroyed by acid rain. Fish and marine mammals in the Gulf were poisoned by south-flowing oil slicks. Russian scientists complained that acidic black rain had fallen at "unprecedented levels" over the former Soviet Union's southern republics. Kerosene falling on northern India triggered a massive algae bloom in Srinagar's Dal Lake.

Climate modelers were not surprised when a typhoon struck Bangladesh on April 30, 1991. Torrential rains driven by 150 mile-per-hour winds lashed the Bangladesh coast and brought 20-foot tidal waves surging over the country's low-lying delta. When the unprecedented rains receded, as many as 200,000 people were dead; at least one million were homeless. Record-breaking floods along China's Yangtze and Huai rivers drove two million from their homes.

While no scientists were willing to link elevated atmospheric soot levels with seasonal storms, monsoon expert Tiruvalam Krishnamurti blamed the intensity of the Bangladesh typhoon on the Gulf oil fires. Iran, Japan, Canada, Britain, and Europe also experienced record-breaking rainfalls under the smoke plume that girdled the globe.

Already destabilized by massive greenhouse emissions, the global atmosphere went into a spasm. In mid-May, California received snows down to the 3,000-foot level five days in succession. Temperatures in Antarctica fell to a historic -107° Fahrenheit. On June 19, 1991, the Northern Hemisphere jet stream vanished. Later that month, the waters off eastern Canada iced over, halting all fishing. Dust storms in India collapsed homes. Beijing was hidden by sand blowing off the Gobi Desert. That winter, bizarre blizzards never before seen by any Middle East inhabitants struck Israel, Jordan, Lebanon, and Syria.

Burning Oil to Control Oil

Ironically, the war that was fought to secure access to cheap oil resulted not only in overwhelming environmental damage caused by the resource it was supposed to protect, but it also unleashed an orgy of oil consumption. During Operation Desert Storm, it took more than 600,000 gallons a day to supply a single armored division of 348 tanks. An F-15 in full afterburner mode devours 240 gallons of fuel per minute. At the height of the conflict, there were 300 U.S. jets stationed on four aircraft carrier groups in the Gulf. Another 700 coalition planes were stationed in Saudi Arabia—including 22 stealth bombers, which use ozone-destroying chemicals to cool their jets' exhausts, thereby masking their infrared "signature." Streaming to the Persian Gulf in just fourteen days, the carrier

USS Independence consumed more than two million gallons of fuel. The 46-day war to protect oil supplies consumed nearly a billion gallons of oil—20 million gallons each day.

In the first two weeks of Operation Desert Shield, more than two billion pounds of weapons, food, medical supplies, and ammunition were trucked from around the United States and shipped by air and sea more than 7,000 miles to the other side of the planet. The volume exceeded the entire 1948 Berlin Airlift.

At least 10 million gallons of sewage were produced every day by the coalition forces. According to a Science for Peace report, coalition forces left behind between 45 and 54 million gallons of sewage in sand pits dug into a desert where nothing decays. This huge lake of effluent posed a threat to Saudi agriculture, as well as animals and humans. Coalition troops also dumped great quantities of solvents, paints, acids, lubricants, and other toxic materials into porous desert sands, along with spilled fuel and discarded explosives.

No allied government—Arab or Occidental—was prepared to pay for the cleanup whose demands on fiscal, human, and logistical resources rivaled the effort to create the mess in the first place. The $100 billion Gulf War did, however, achieve its main objective of securing an important source of fuel to supply America's oil addiction. It also established the first permanent U.S. bases in the Middle East, while opening the way for U.S. drilling rights in Kuwait's formerly exclusive oil fields.

The war's biggest lesson—that oil dependency can lead to truly staggering costs—was lost on the Bush administration, which announced its new oil strategy during the desert war. The long-awaited energy plan stressed oil drilling offshore and in Alaska's Arctic National Wildlife Refuge. As if to drive home its anti-ecological stance, the Bush administration also used the Gulf crisis as an excuse for exempting the military from environmental impact studies.

Serbia: The Impact of NATO's Bombs

Phillip Frazer (2000)

NATO'S bombs not only destroyed Serbia's military machine during the alliance's 1999 intervention; they also devastated the region's land, air, and water. After several months of Serbian forces burning villages and NATO flying over 40,000 sorties that dropped powerful explosives on Yugoslavia, massive damage had been inflicted on the Yugoslav environment and neighboring countries. Western estimates are that property damage from NATO's bombs exceeded $100 billion.

Yugoslavia has a network of national parks—some of which it claims NATO "intensively bombed"—and is home to some 250 rare native plants and more than 420 rare native animal species. The country boasts its own environmental movement. Montenegro, Serbia's junior partner in the Yugoslav Federation, had declared itself the world's first "environmental state," pledging to live in harmony with nature.

The most apparent environmental catastrophes were NATO's repeated bombings of the Petrohemija petroleum processing plant and a chemical and fertilizer plant in Pancevo, a city of 140,000.

The *London Times* reported that the bombing and fires released "a toxic cloud of smoke and gas hundreds of feet high," containing hydrochloric acid, toxic phosgene, and chlorine gases. Concentrations of vinyl chloride reached 0.43 milligrams per cubic centimeter—8,600 times higher than the danger level. Workers at the complex released 1,400 tons of carcinogenic ethylene dichloride and 800 tons of hydrogen chloride into the Danube to avoid the risk of a further explosion. The pollution was left to flow downstream to Rumania and Bulgaria and into the Black Sea.

The Serbian newspaper *Glas Javnosti* reported that the bombing of a chlorine-alkaline electrolysis plant at Pancevo released "3,000 tons of [sodium] hydroxide, 30 tons of liquid chlorine and almost 100 tons of mercury." The Yugoslav government accused NATO of "ecocide," claiming that had the chlorine and ammonia tanks not been emptied before the bombing, the attack "would have sounded the death knell for all the inhabitants of Pancevo and its surroundings."

People exposed to the cloud have already started dying. Children with chemical-singed lungs found it painful to breathe and an unusual number of women suffered miscarriages following the bombing. Milan Borna, chief of Pancevo's environmental protection department told the Associated Press: "We fear that the worst effects may be degenerative changes in future generations."

Robert Hayden, director of the Center for Russian and East European Studies at the University of Pittsburgh, said: "NATO's bombing of the petrochemical plant . . . risked the life, health and safety of the civilian population of 2 million in the city of Belgrade."

According to Green Cross, a global environmental protection organization headed by former USSR president Mikhail Gorbachev, "about 1,200 tons of vinyl-chloride monomers was released into the air." When ignited, this colorless, flammable, toxic gas produces hydrogen chloride, carbon dioxide, carbon monoxide, chlorine, and small amounts of deadly phosgene nerve gas.

An attack on a transformer station at Pancevo released stockpiles of polychlorinated biphenyls (PCBs), cancer-causing compounds that were banned worldwide decades ago. Just one liter of PCBs can contaminate one billion liters [264 million gallons] of water to levels dangerous to human health. According to Yugoslavian authorities, "several tons of PCBs [were] released into the Danube River watershed."

Many oil refineries and storage facilities were bombed, including one at Novi Sad in the northern Vojvodina region, which lies on the banks of the Danube. The bombings produced an oil spill in the Danube nine miles long. During the bombing campaign, the *Glasgow Herald* warned of the possibility that the oil slick could jam the machines that pump water from the Danube to the reactor cooling system of Bulgaria's Kozloduy nuclear power plant, running the risk of triggering "a second Chernobyl." In an interview with the BBC, Yugoslavian hydrogeologist Momir Komatina worried that the region's underground water resources—which serve about 90 percent of Serbia's domestic and industrial needs—are now at risk. He noted that the imperiled water tables extend well beyond Serbia and into southern Europe. Polluted surface waters eventually will make their way to the Aegean and Adriatic Seas and then into the Mediterranean.

Blowing in the Wind

In addition to the Serbs burning the homes of ethnic Albanians, the explosion of 23,000 NATO bombs (more than 14 million pounds of explosives) and the destruction of chemical facilities, refineries, fuel tanks, and warehouses sent a significant volume of greenhouse gases and vast clouds of toxic soot into the atmosphere. The night after the bombing of oil depots near the Yugoslav-Bulgarian border, heavy rains blackened with soot fell on Bulgaria.

The Macedonian Environment Ministry's monitoring stations detected the presence of airborne furans and dioxins—toxic and carcinogenic substances. The ministry assumes the pollutants were blown over Macedonia after NATO bombs hit refineries and transformer stations.

During the thirty-two-day air assault, vast quantities of jet engine exhaust were dumped into the regional airspace. More than 40,000 flights by high-flying NATO aircraft released chemically reactive exhaust trails filled with nitrogen oxides that devour stratospheric ozone.

Jet aircraft emissions are far more toxic than the exhaust fumes of gasoline or even industrial fuels. The toxic and carcinogenic combustion products can destroy forests and vegetation, and impair human health as well. The combustion of fuel stabilizers with lead and fluorine compounds produces fluorine radicals that can cause painful wounds and burns in humans and wildlife. Exhaust gases eventually drift to earth—all the way from the launch site to the target.

The burning of newer buildings—such as the TV headquarters and government office blocks bombed in downtown Belgrade—produced fumes that were far more toxic than smoke from pre–World War II structures.

Radioactive Shells

One controversy that got occasional coverage in the U.S. media was NATO's use of tank-piercing depleted uranium (DU) munitions. DU is a very dense metal, capable of penetrating steel plating up to fifty-seven millimeters thick. It is also toxic, carcinogenic, and still 60 percent as radioactive as natural uranium. The Iraqis blame DU for a range of health problems in southern Iraq where it was used in the Gulf War. In the United States, some veterans hold DU responsible for causing Gulf War Syndrome, whose symptoms resemble radiation sickness.

When Belgrade broadcasts first announced that NATO was using "radioactive weapons . . . which are forbidden by the Geneva Convention," *ABC Nightline* dismissed the charge as an "astonishing claim." Asked the next morning about NATO's use of DU weapons, Pentagon spokesperson Kenneth Bacon replied that such discussions were "verboten from this podium."

John Catalinotto, an editor of the 1997 book *Metal of Dishonor: Depleted Uranium*, claims that just one "hot particle" of DU in the lungs is equivalent to receiving a chest X-ray every hour. Once inhaled or ingested, DU is impossible to remove, so the lung is irradiated at that rate for life—whether the victim is a Yugoslav citizen or a NATO soldier.

In June 1999, Russian leader Mikhail Gorbachev charged that weapons containing depleted uranium "burn at high temperatures, producing poisonous clouds of uranium oxide that dissolve in the pulmonary and bronchial fluids. Anyone within the radius of 300 meters from the epicenter of the explosion inhales large amounts of such particles."

In the aftermath of the war, Gorbachev called for a global ban on military strikes against nuclear power stations and chemical and petrochemical plants. "Similarly," Gorbachev argues, "we should prohibit weapons whose use may have particularly dangerous, long-term, and massive environmental and medical consequences. In my view, weapons containing depleted uranium should be among the first considered for such a ban."

Gorbachev called for a conference on the environmental consequences of war to set rules for future conflicts. The first conference was held in Washington in 1998.

Waging War on Global Laws

On December 18, 1998, when the United States and Britain began bombing urban centers in Iraq, then-secretary of defense William Cohen told

NBC's *Today* program, "We're not going to take a chance and try to target any facility that would release any kind of horrific damage to innocent people." Three months later, NATO and the United States began targeting industrial sites in urban settings. US/NATO also began specifically targeting civilian telephone and computer sites, political offices, and the homes of Serbian leaders. A NATO attack on Yugoslavia's TV headquarters in Belgrade killed a number of civilian journalists.

NATO's bombs also damaged the Serbian Historical Museum, the Church of Saints Peter and Paul, the 400-year-old Rakovica Monastery, the Children's Culture Center, and the Dusko Rodovic Theater.

On May 11, 1999, Human Rights Watch (HRW) appealed to NATO to halt the use of antipersonnel cluster bombs. Because the brightly colored bomblets resemble soda cans and baseballs, they frequently attract—and kill—children. "The U.S. may not have signed the landmines treaty, but it's still obliged to carry out warfare according to international humanitarian law," declared HRW's Joost Hiltermann.

An agronomist with the New Green Party in Belgrade noted that NATO's bombing would prevent "the planting of 2.5 million hectares of land. . . . The lack of fuel for agricultural machines will have catastrophic results, because it leads to hunger of the entire population. When you add to this the poisoning of the water, air, and soil, catastrophe becomes a cataclysm."

Thousands of citizens from hundreds of nations have taken years to write agreements designed to slow or stop dangerous trends such as the destruction of the ozone layer. One military decision by a politician can set such efforts back decades. The Yugoslavian Ministry for the Environment claims NATO violated just about every existing environmental treaty, including the Rio Declaration on the Environment and Development.

Even during war, both NATO and Yugoslavian forces must abide by international laws created to protect civilians and the environment. NATO's actions routinely ignored agreements to protect the ozone layer; slow the greenhouse effect; preserve wild flora and fauna and their natural habitats; protect biological diversity; deal with the effects of industrial accidents and dangerous wastes; protect waterways; prevent pollution from shipping vessels; and protect cultural monuments and natural treasures.

The 1977 UN Treaty against Environmental Modification prohibits widespread modification of the environment, but it does not go far enough. An international campaign is under way to counter the "new generation of environmental weapons" unleashed by NATO during the Serb war. There is a desperate need to address the environmental damage of the NATO attacks. Meanwhile, the *Christian Science Monitor* notes, "The U.S. and several key European allies have indicated that, other than the most basic humanitarian aid, no international assistance will be granted to Serbia as long as Milosevic, an indicted war criminal, is in power." The White House needs to be reminded that environmental restoration is a basic human right.

Sardinia: Bombs and Cancer in Paradise

Helen Jaccard, Veterans for Peace (2012)

The sound of bombs, missiles, and explosions; massive attacks from the sea onto the beach; an epidemic of cancers and birth defects; soil, air, food, and water contaminated with heavy metals, jet fuel, napalm, and other poisons. Is this a modern war zone? No—Sardinia is the victim of weapons manufacturers, polluting military activities, and a political system that cares about power and money over the health of people and the environment.

Sardinia is an island paradise in the Mediterranean Sea with diverse wildlife and beautiful beaches. The shepherds and farmers produce magnificent wine, honey, and cheese. "We are peaceful people, poor ones maybe, but very welcoming," says Sardinian environmental engineer Alice Scanu. "That's how I'd like Sardinians to be remembered: not as people involved in wars and power games."

For more than fifty years, Sardinia has been used to train soldiers and pilots to master various war scenarios while launching bombing sorties and testing new bullets, bombs, missiles, and drones. Seventy percent of Italy's military bases are located in Sardinia, where Italian and NATO bases occupy about one-third of the island's acres. During military exercises, about 7,200 square miles are closed to public navigation and fishing. During all of June 2014, the military conducted war practice on the "protected" Lake Omodeo over the objection of the people.

The bases generate large quantities of wastes containing antimony, cadmium, cerium, lead, and thorium, contaminating beaches and water. Explosions of weapons and waste from past wars leave vast areas—up to 2,000 square meters each—unable to support vegetation. Each explosion can produce as much pollution as a municipal solid waste incinerator generates in a year, exposing downwind communities—shepherds, soldiers, and animals alike—to clouds of toxic dust.

Thorium—a radioactive and highly carcinogenic heavy metal—has been found in Sardinian honey and milk. Pieces of bombs, missiles, and bullets can be found lying on the ground and abandoned in the sea. Unexploded ordnance lies in and around the restricted areas on both land and water.

High rates of cancer and birth deformities are found near the firing ranges where soldiers launch artillery rockets, drones, and laser-guided precision bombs (some of which contain depleted uranium, asbestos, and white phosphorus).

In one village of 150 inhabitants, there had been *no* cases of leukemia or lymphoma before 2000. In 2002, however, twelve people died from leukemia, and another sixty-three leukemia deaths occurred over the next ten years. Serious cancers were found in 65 percent of workers on seven farms located near one base while rates of lymphoma, thyroid cancer, and autoimmune diseases also were unusually high.

Evandro Lodi Rizzini, a nuclear physicist with CERN (European Organization for Nuclear Research), has found elevated levels of radioactive thorium-232 and cerium (proving that the thorium was man-made) in tissue samples taken from fifteen of eighteen local shepherds who died of cancer between 1995 and 2000.

Between 1988 and 2002, fourteen children were born with severe malformations in a small village bordering one base. Malformed animals are born near the military bases—including two-headed lambs, calves with deformed legs, and a pig with one huge, grotesque eye. A tissue sample from one malformed lamb was found to contain depleted uranium.

John Madeddu worked at Sardinia's Capo Frasca base from 1968 to 1987 and now suffers from diffuse large-cell lymphoma. He remembers a clearing inside the base where a large number of bullets had accumulated. When rains flooded the area, the water seeped into the ground, infiltrating the artesian wells that provide water for both the base and nearby farms. No cleanup efforts were ever undertaken. Sheep still graze on this contaminated land (and local farmers still sell mutton and cheese for a living). Cattle still graze near the bases. Even if hit and killed by bullets or metal fragments containing heavy metals, these animals are still butchered and eaten.

Southwest Sardinia is home to Decimomannu, NATO's largest air base, used since 1954 by Italy, Germany, Canada, and the United States. The air base has contaminated the land and water with jet fuel containing carcinogenic xylene, benzene, and lead. In February 2011, Mayor Louis Porceddu was forced to prohibit the use of water drawn from local wells.

A local shepherd analyzed the situation with clear, shocking realism: "I have leukemia, I have only a few months or years of life, I accepted it. Nobody cares about us, and we just do not count for anything. They are powerful; it is better for them if there are fewer of us."

The Nuking of La Maddalena

La Maddalena is an archipelago located northeast of Sardinia that draws tens of thousands of tourists who come to enjoy the beautiful beaches and hiking trails. Until 2008, La Maddalena also played host to a U.S./NATO submarine base on Santo Stefano Island. In 2003, the nuclear-powered submarine *USS Hartford* struck a rock, damaging its rudders. However, the island's residents suspect that even greater damage was done. Their concerns increased after a team lead by Massimo Zucchetti, a professor at the Torino Polytechnic, analyzed algae in the archipelago and found traces of radioactive alpha particles and plutonium, sometimes in high concentrations.

Dr. Antonietta Gatti, an experimental physicist at the University of Bologna, has found nanoparticles of iron, lead, tungsten, and copper in the tissues of citizens and sheep. "Rain leads to the contamination of the soil," she explained. "Through air pollution, other areas that are not involved in the testing are contaminated as well."

"Here in Sardinia, we are confronted with war victims—but in a peaceful area," says Giancarlo Piras, a Sardinia resident. "We like to call this area the 'Zone for Preparing New Wars.'" Under existing law, military officials are supposed to tell local governments what kinds of weapons and materials are being tested at the bases. The reality is that none of the armies provide this information, hiding under the umbrella of "military secrecy."

In the 1990s, local fishers who had been driven from their profession by NATO naval exercises became activists, demanding their traditional right to fish in the sea. Acts of civil disobedience were staged—at the port and base entrances and at sea. The fishers boldly challenged the restrictions by directing more than forty boats into the heart of NATO's war games and throwing their fishing nets into a "prohibited" area of ocean claimed by the war ships.

Their demands were simple: the right to safe work and to have their stolen sea returned. The protests only ended in 2005 when the military agreed to pay the fishers to stay out of the water. Many of the fishermen have since abandoned their nets.

Other protests continue. On the fifteenth of each month in Cagliari, cancer victims, their families, and others rally against the military use of Sardinia. The protesters have five clear demands: (1) Reveal what chemicals and metals have been spilled. (2) Decontaminate the surrounding land, aquifers, and sea. (3) Provide health care to everyone harmed by military activity. (4) Provide financial aid as well as clean land and sea to farmers and fishers. (5) Close the bases and demilitarize the island.

Vietnam: Delivering Death to the A Luoi Valley

Dr. Elizabeth Kemf (1990)

In the autumn of 1985, ten years after the U.S.-Vietnam War ended, I traveled to Vietnam for the first time to observe the long-term environmental effects of that war. As an American citizen and as a naturalist, I wanted to know how the landscape and the people had been changed by the spraying of 72 million liters [19 million gallons] of herbicides and the displacement of billions of cubic yards of soil by 13 million tons of bombs. Reports indicated that the country was pocked with some 25 million bomb craters and that some 2.2 million hectares [5.4 million acres] of forest had been lost or affected as a direct result of intensive bombing, napalm raids, and the massive mechanized land clearing of tropical forests with the world's largest bulldozers. I wanted to see if the legacy of "ecocide" was still evident nearly twenty years later.

While riding along Highway Nine toward the Laotian border, my guide, Mr. Tuy (the director of the Provincial Forestry Department), pointed out land not yet cleared of unexploded munitions. These untouched patches increased in number as we approached Khe Sanh, one of the bloodiest battlefields of the Vietnam War. The roadside was flanked with high clumps of "pernicious grass." Once a thick tropical forest, the terrain is now denuded of trees and the coarse grass has taken over. Mr. Tuy said that when he first visited the area in 1977, nothing remained but the stumps of charred trees.

A swath of tropical forest and mountains covering some 124,000 hectares has been largely reduced to degraded grassland. Eighty percent of the forest is gone. According to *Operation Ranch Hand*, a report issued by the U.S. Air Force in 1982, "by June 30, 1966, Ranch Hand had sprayed approximately 1,500 kilometers [932 miles] of roads and trails to a depth of 250 meters [820 feet] on each side—the result of 200 sorties and 200,000 gallons of herbicides."

As we crossed the Da Krong River and began our entry into the A Luoi Valley, it was as if someone had waved a magic wand and transformed the wasteland into a lush jungle. The forest was virgin again, revealing itself in a sudden burst of splendor. The American C-123 cargo planes that used to spray the trees and crops had not defiled this stretch of woods. Tall luxuriant trees covered the mountainside in canopied layers of variegated green. Towering twenty to thirty meters (sixty-six to ninety-eight feet) in height, the thick and damp trees stood in several layers of dense vegetation—the heart of one of Vietnam's last primary forests.

Plague is a growing health problem in areas sprayed by Agent Orange because of the proliferation of grass, providing ideal conditions for the multiplication of rats. When we reached the area destroyed by Agent

Orange, there was no need to ask if we had arrived. On both sides of the valley, the hills were stripped bare, their ridges lined with the fragmented trunks of dead trees. As we drove for miles through the stark countryside, I was reminded of the words of Rachel Carson who, more than a quarter-century ago, sounded the first warning against the use of pesticides. The valley was having a "silent spring"—no birds, no people, no wildlife, just miles of devastated landscape.

I was speechless. This stretch of A Luoi Valley, some 60 km long, represented 100,000 hectares of forest destroyed by Agent Orange—yet it amounts to less than 5 percent of the overall damage wreaked by herbicides and bombs and indiscriminate plowing up of forest and field in the Vietnamese countryside. This was a crime against nature and humanity, and it was beyond retribution. The rich fauna, the large number of species supported by the forest, are gone, and the desperately poor tribal people have been deprived of their economic base.

Mr. Xuan Hien, secretary of the People's Committee, received us at the forestry station. He related the war history of the area:

"Spraying started in 1965 and was stepped up to eight spray missions a year during 1968 to 1969," he said. "I believe that my four children all died as a result of the spraying. . . . All the inhabitants here were affected by the spraying and the bombing. They had to go underground into small A-shaped trenches. The people built covers for the trenches and pulled these on top of the holes to protect themselves until the planes stopped spraying and bombing.

"The smell was terrible. Many of the people died. Some parts of people's bodies just shriveled up. Some were paralyzed and are still affected. Farm animals died. Almost all wild animals in the forest disappeared. We found the dead carcasses of birds, tigers, and monkeys after the spraying. We used to rely on the forest for our food. We had nothing left, no forest, no fruit, no chickens, no buffalo, nothing to hunt. Our crops died, even the fish in the river died, or were so strange-looking, we didn't dare eat them."

When we arrived at the site, which was not only sprayed by Agent Orange but also bombarded and shelled by both U.S. and North Vietnamese troops, the director explained that one of Vietnam's richest inland tropical forests once covered the land.

I had the impression we had wandered into the world's largest junkyard. Tanks, guns, sandbags, and detonated rockets were strewn across the landscape as far as the eye could see. Heaps of rusting metal were piled on the main path around which grew acres of pernicious grass several feet high. The young boys who were not lugging metal scraps toward one of the scrap heaps were digging rubble from enormous bomb craters, many more than 10 meters wide.

"This whole forest was full of wild animals and birds. Our forests are very rich," he said. "In the last seven years we planted, together with the Con Tien secondary school, over one million trees."

Speaking a quarter of a century ago, Ho Chi Minh inspired the country with a statement that is still quoted by ecologists throughout Vietnam: "Forest is gold. If we know how to conserve and use it well, it will be very precious."

COLLATERAL DAMAGE

Every creature is better alive than dead, men and moose and pine trees, and he who understands it aright will rather preserve its life than destroy it.

—Henry David Thoreau, essayist, abolitionist, naturalist, and war-tax resister

Baghdad: A Civilization Torn to Pieces

Robert Fisk, The Independent (2003)

BAGHDAD (April 13, 2003)—They lie across the floor in tens of thousands of pieces, the priceless antiquities of Iraq's history. The looters had gone from shelf to shelf, systematically pulling down the statues and pots and amphorae of the Assyrians and the Babylonians, the Sumerians, the Medes, the Persians, and the Greeks and hurling them on to the concrete.

Our feet crunched on the wreckage of 5,000-year-old marble plinths and stone statuary and pots that had endured every siege of Baghdad, every invasion of Iraq throughout history—only to be destroyed when America came to "liberate" the city. The Iraqis did it. They did it to their own history, physically destroying the evidence of their own nation's thousands of years of civilization.

Not since the Taliban embarked on their orgy of destruction against the Buddhas of Bamiyan and the statues in the museum of Kabul—perhaps not since the Second World War or earlier—have so many archaeological treasures been wantonly and systematically smashed to pieces.

"This is what our own people did to their history," the man in the gray gown said as we flicked our torches yesterday across the piles of once perfect Sumerian pots and Greek statues, now headless, armless, in the storeroom of Iraq's National Archaeological Museum. "We need the American soldiers to guard what we have left. We need the Americans here. We need policemen." But all that the museum guard, Abdul-Setar Abdul-Jaber, experienced yesterday was gun battles between looters and local residents, the bullets hissing over our heads outside the museum and skittering up the walls of neighboring apartment blocks.

"Look at this," he said, picking up a massive hunk of pottery, its delicate patterns and beautifully decorated lips coming to a sudden end where the jar—perhaps two-feet-high in its original form—had been smashed into four pieces. "This was Assyrian." The Assyrians ruled almost 2,000 years before Christ.

And what were the Americans doing as the new rulers of Baghdad? Why, yesterday morning they were recruiting Saddam Hussein's hated former policemen to restore law and order on their behalf. (The last army to do anything like this was [Lord Louis] Mountbatten's force in Southeast Asia, which employed the defeated Japanese army to control the streets of Saigon—with their bayonets fixed—after the recapture of Indo-China in 1945.)

But "liberation" has already turned into occupation. Faced by a crowd of angry Iraqis in Firdos Square demanding a new Iraqi government "for our protection and security and peace," U.S. Marines, who should have been providing that protection, stood shoulder to shoulder facing them, guns at the ready. The reality, which the Americans—and, of course, Mr. [U.S. defense secretary Donald] Rumsfeld—fail to understand is that, under Saddam Hussein, the poor and deprived were always the Shia Muslims, the middle classes always the Sunnis, just as Saddam himself was a Sunni. So it is the Sunnis who are now suffering plunder at the hands of the Shia.

And so the gun-fighting that broke out yesterday between property owners and looters was, in effect, a conflict between Sunni and Shia

Muslims. By failing to end this violence—by stoking ethnic hatred through their inactivity—the Americans are now provoking a civil war in Baghdad.

Yesterday evening, I drove through the city for more than an hour. Hundreds of streets are now barricaded with breezeblocks, burnt cars, and tree trunks, watched over by armed men who are ready to kill strangers who threaten their homes or shops. Which is just how the civil war began in Beirut in 1975.

A few U.S. Marine patrols did dare to venture into the suburbs yesterday—positioning themselves next to hospitals which had already been looted—but fires burnt across the city at dusk for the third consecutive day. The municipality building was blazing away last night, and on the horizon other great fires were sending columns of smoke miles high into the air.

No Power, No Water, No Law, No Order

Too little, too late. Yesterday, a group of chemical engineers and water purification workers turned up at the U.S. Marine headquarters, pleading for protection so they could return to their jobs. Electrical supply workers came along, too. But Baghdad is already a city at war with itself, at the mercy of gunmen and thieves.

There is no electricity in Baghdad—as there is no water and no law and no order—and so we stumbled in the darkness of the museum basement, tripping over toppled statues and stumbling into broken winged bulls. When I shone my torch over one far shelf, I drew in my breath. Every pot and jar—"3,500 BC" it said on one shelf corner—had been bashed to pieces.

Why? How could they do this? Why, when the city was already burning, when anarchy had been let loose and less than three months after U.S. archaeologists and Pentagon officials met to discuss the country's treasures and put the Baghdad Archaeological Museum on a military database—did the Americans allow the mobs to destroy the priceless heritage of ancient Mesopotamia?

For well over 200 years, Western and local archaeologists have gathered up the remnants of this center of early civilization from palaces, ziggurats, and 3,000-year-old graves. Their tens of thousands of handwritten card index files—often in English and in graceful nineteenth-century handwriting—now lie strewn amid the broken statuary. I picked up a tiny shard. "Late 2nd century, no. 1680" was written in pencil on the inside.

To reach the storeroom, the mobs had broken through massive steel doors, entering from a back courtyard and heaving statues and treasures to cars and trucks.

The looters had left only a few hours before I arrived and no one—not even the museum guard in the gray gown—had any idea how much they had taken. A glass case that had once held 40,000-year-old stone and flint objects had been smashed open. It lay empty. No one knows what happened to the Assyrian reliefs from the royal palace of Khorsabad, nor the 5,000-year-old seals nor the 4,500-year-old gold leaf earrings once buried with Sumerian princesses. It will take decades to sort through what they have left, the broken stone torsos, the tomb treasures, the bits of jewelry glinting amid the piles of smashed pots.

The mobs that came here—Shia Muslims, for the most part, from the hovels of Saddam City—probably had no idea of the value of the pots or statues. Their destruction appears to have been the result of ignorance as much as fury. In the vast museum library, only a few books—mostly mid-nineteenth-century archaeological works—appeared to have been stolen or destroyed. Looters set little value in books.

I found a complete set of the *Geographical Journal* from 1893 to 1936 still intact—lying next to them was a paperback titled *Baghdad, The City of Peace*—but thousands of card index sheets had been flung from their boxes over stairwells and banisters.

Even as the Americans encircled Baghdad, Saddam Hussein's soldiers showed almost the same contempt for its treasures as the looters. Their slit trenches and empty artillery positions are still clearly visible in the museum lawns, one of them dug beside a huge stone statue of a winged bull.

Only a few weeks ago, Jabir Khalil Ibrahim, the director of Iraq's State Board of Antiquities, referred to the museum's contents as "the heritage of the nation." They were, he said, "not just things to see and enjoy. We get strength from them to look to the future. They represent the glory of Iraq."

Mr. Ibrahim has vanished, like so many government employees in Baghdad, and Mr. Abdul-Jaber and his colleagues are now trying to defend what is left of the country's history with a collection of Kalashnikov rifles.

"We don't want to have guns, but everyone must have them now," he told me. "We have to defend ourselves because the Americans have let this happen. They made a war against one man—so why do they abandon us to this war and these criminals?"

Half an hour later, I contacted the civil affairs unit of the U.S. Marines in Saadun Street and gave them the exact location of the museum and the condition of its contents. A captain told me that "we're probably going to get down there." Too late. Iraq's history had already been trashed by the looters whom the Americans unleashed on the city during their "liberation."

Priceless Documents Set Ablaze

BAGHDAD (April 15, 2003)—So yesterday was the burning of books. First came the looters, then the arsonists. It was the final chapter in the sacking of Baghdad. The National Library and Archives—a priceless treasure of Ottoman historical documents, including the old royal archives of Iraq—were turned to ashes in 3,000 degrees of heat. Then the library of Korans at the Ministry of Religious Endowment was set ablaze.

I saw the looters. One of them cursed me when I tried to reclaim a book of Islamic law from a boy of no more than ten. Amid the ashes of Iraqi history, I found a file blowing in the wind outside: pages of handwritten letters between the court of Sharif Hussein of Mecca, who started the Arab revolt against the Turks for Lawrence of Arabia, and the Ottoman rulers of Baghdad.

And the Americans did nothing. All over the filthy yard they blew, letters of recommendation to the courts of Arabia, demands for ammunition for troops, reports on the theft of camels and attacks on pilgrims, all in delicate handwritten Arabic script. I was holding in my hands the last Baghdad vestiges of Iraq's written history. But for Iraq, this is Year Zero; with the destruction of the antiquities in the Museum of Archaeology on Saturday and the burning of the National Archives and then the Koranic library, the cultural identity of Iraq is being erased. Why? Who set these fires? For what insane purpose is this heritage being destroyed?

When I caught sight of the Koranic library burning—flames 100 feet high were bursting from the windows—I raced to the offices of the occupying power, the U.S. Marines' Civil Affairs Bureau. An officer shouted to a colleague that "this guy says some biblical [*sic*] library is on fire." I gave the map location, the precise name—in Arabic and English. I said the smoke could be seen from three miles away and it would take only five minutes to drive there. Half an hour later, there wasn't an American at the scene—and the flames were shooting 200 feet into the air.

There was a time when the Arabs said that their books were written in Cairo, printed in Beirut, and read in Baghdad. Now they burn libraries in Baghdad. In the National Archives were not just the Ottoman records of the caliphate, but even the dark years of the country's modern history, handwritten accounts of the 1980–88 Iran-Iraq war, with personal photographs and military diaries, and microfiche copies of Arabic newspapers going back to the early 1900s.

But the older files and archives were on the upper floors of the library where petrol must have been used to set fire so expertly to the building. The heat was such that the marble flooring had buckled upwards and the concrete stairs that I climbed had been cracked.

The papers on the floor were almost too hot to touch, bore no print or writing, and crumbled into ash the moment I picked them up. Again, standing in this shroud of blue smoke and embers, I asked the same question: Why?

So, as an all-too-painful reflection on what this means, let me quote from the shreds of paper that I found on the road outside, blowing in the wind, written by long-dead men who wrote to the Sublime Porte in Istanbul or to the Court of Sharif of Mecca with expressions of loyalty and who signed themselves "your slave." There was a request to protect a camel convoy of tea, rice, and sugar, signed by Husni Attiya al-Hijazi (recommending Abdul Ghani-Naim and Ahmed Kindi as honest merchants), and a request for perfume and advice from Jaber al-Ayashi of the royal court of Sharif Hussein to Baghdad to warn of robbers in the desert. "This is just to give you our advice for which you will be highly rewarded," Ayashi says. "If you don't take our advice, then we have warned you." A touch of Saddam there, I thought. The date was 1912.

Some of the documents list the cost of bullets, military horses, and artillery for Ottoman armies in Baghdad and Arabia; others record the opening of the first telephone exchange in the Hejaz—soon to be Saudi Arabia—while one recounts, from the village of Azrak in modern-day

Jordan, the theft of clothes from a camel train by Ali bin Kassem, who attacked his interrogators "with a knife and tried to stab them but was restrained and later bought off." There is a nineteenth-century letter of recommendation for a merchant, Yahyia Messoudi, "a man of the highest morals, of good conduct and who works with the [Ottoman] government." This, in other words, was the tapestry of Arab history—all that is left of it, which fell into *The Independent*'s hands as the mass of documents crackled in the immense heat of the ruins.

For almost a thousand years, Baghdad was the cultural capital of the Arab world, the most literate population in the Middle East. Genghis Khan's grandson burnt the city in the thirteenth century and, so it was said, the Tigris River ran black with the ink of books. Yesterday, the black ashes of thousands of ancient documents filled the skies of Iraq. Why?

Ukraine: Civil War and Combat Pollution

Doug Weir, Toxic Remnants of War Project (2016)

Clean air, water, and food are essential to survival, therefore civilian protection during and after armed conflict requires the effective protection of the environment. There is growing recognition by states, militaries, and international organizations of the polluting impact of conflict and military practices. The term "toxic remnants of war" (TRW) has been coined to facilitate greater scrutiny of the subject.

TRW can be defined as: "Any toxic or radiological substance resulting from military activities that forms a hazard to humans and ecosystems." "Direct TRW" are an immediate result of military activity. For example, pollution released from attacking industrial targets or toxic residue resulting from munitions use. "Indirect TRW" result from events or conditions connected to conflict. For example, during the 2003 invasion of Iraq a number of industrial sites were damaged by conflict or simply abandoned. These derelict and unsecured sites were subsequently looted, exposing people to highly toxic chemical and radioactive substances.

The ongoing war in Ukraine has generated both direct and indirect TRW impacts.

Soon after violent secessionist conflict erupted in eastern Ukraine in 2014, information began to emerge on the environmental impacts of war across the highly industrialized Donbas region. Although obtaining accurate data was difficult, indications were that the conflict had resulted in potentially long-term damage to the environment that posed a number of civilian health risks.

The environmental legacy of conflict and military activities is rarely prioritized in post-conflict response, in spite of the short- and long-term impacts on civilian health and livelihoods. Warfare in highly industrialized areas has the potential to generate new pollution incidents and exacerbate existing problems. The conflict in Ukraine has done both, as well as damage the area's natural environment.

With the signing of the second round of Minsk agreements in February 2015, hope re-emerged that a peaceful solution might be possible. But should the fragile truce collapse, an enlarged conflict would pose new and grave risks to the region's people and environment.

Prior to the outbreak of the war, more than 5,300 industrial enterprises were operating in the prewar Donetsk and Luhansk oblasts (provinces). Initial war damage to the region's industry has been widespread, ranging from direct damage to industrial installations to enterprises simply stopping production because of the lack of raw materials, energy, workforce, or distribution channels.

In some cases, the disruption led to accidental releases of pollutants from shelled or bombed facilities. In others, facilities have been forced to shift to more polluting technologies that have compromised regional air quality. Among the dozens of facilities damaged by fighting were the Zasyadko coal mine, a chemicals depot at Yasynivskyi coke and chemical works in Makiyvka, the Lysychyansk oil refinery, an explosives factory at Petrovske, and a fuel-oil storage facility at Slavyansk thermal power plant.

Coal mining has been the backbone of the economy of the Donbas region since the nineteenth century. With the intermittent disruption of the electricity supply across the conflict area, ventilation systems and water pumps in coal mines failed, resulting in flooding and the release of accumulated gases after ventilation restarted. The often irreparable flooding not only damages the mines but also waterlogs adjacent areas and pollutes groundwater. By April 2015, permanent or temporary flooding resulting from the military conflict had been reported at more than ten mines.

The Zasyadko mine in Donetsk used to produce four million tons of coal annually and was one of the region's economic flagships. An explosive release of methane in March 2015 killed 33 of the 200 miners underground at the time. Even though this was not the first such accident at Zasyadko (which is ranked among the mining industry's deadliest mines), the chair of the mine's board attributed the incident to heavy shelling at nearby Donetsk Airport, where fighting continued until late January 2015.

There have been numerous media reports about war damage caused to Donbas' water supply, including in and around Luhansk and Donetsk—cities with a combined prewar population of 1.5 million. While repair work to the water infrastructure was carried out—often under direct fire—periods of irregular supply were common. Less well documented is the impact of the conflict on drinking water quality, but one can reasonably assume the disruptions have caused widespread deterioration.

Limited sampling by the Ukraine-based NGO Environment-People-Law confirmed the expected range of some "war chemicals" from the use of conventional weapons in impact zones. Similarly, large quantities of damaged military equipment and potentially hazardous building rubble will require disposal. The Ukrainian Ministry of Defense also raised concerns that depleted uranium weapons may have been used in the fighting around Donetsk Airport, and proposed to determine whether this was the case when conditions allowed.

The conflict has also damaged the region's numerous nature protection areas, from armed groups occupying administrative buildings to the impact of fighting and the movement of heavy vehicles within nature reserves. The restoration of large tracts of agricultural and other land for normal cultivation and use will require considerable effort too, and will be complicated by the presence of new minefields and unexploded ordnance.

As is common for armed conflicts in heavily developed areas, a large proportion of the pollution impact may not come directly from the fighting but from damage to industrial infrastructure and the disruption of everyday economic activities. A good example from the Donbas region was seen in data from its only functioning automated air quality monitoring station, located in the town of Schastya in the Luhansk oblast and which remained in operation until November 2014. The data demonstrate that peak concentrations of airborne contamination are not only associated with periods of combat. In this case, instead, the pollution peaks were linked to a reduction in the supply of high-grade coal for the Luhanska power plant in August 2014.

Coal supplies were first restricted when a bridge in Kondrashevskaya-Novaya was destroyed. Then an electrical substation was shelled, which disconnected the area from the rest of Ukraine's electricity grid. As a result, the Luhanska power plant, which was responsible for supplying more than 90 percent of the oblast's electricity, was forced to simultaneously increase production while turning to lower-grade coal. This caused a clear deterioration in air quality.

It is impossible to predict whether further damage will be wreaked on the people and the environment of Donbas. Insecurity continues to impact basic environmental governance on both sides of the conflict, while cooperation across the front line—even on urgent humanitarian issues—remains a remote prospect.

Because of the great potential for long-term civilian health risks from the conflict pollution, efforts to collect systematic data on pollution and health outcomes should start immediately, as must preparations for remediation. The financial and technical requirements for the comprehensive assessment and remediation of contaminated sites will be considerable.

There is consensus among experts, including the International Committee of the Red Cross (ICRC), that legal protection for the environment during war is inadequate and needs further development. A major limitation of treaty-based international humanitarian law is the high threshold of damage required for it to take effect and the fact that obligations for remediation are not covered. Customary international law is thought to have good potential to address these deficiencies, but the experiences from the bans on land mines and cluster munitions suggest that a new system of environmental protection and restoration may be required.

As the ICRC noted in 2011, we need to consider creating new systems to provide the environmental assistance required to protect both civilians and the environment from conflict pollution: "Given the complexity, for example, of repairing damaged plants and installations or cleaning up polluted soil and rubble, it would also be desirable to develop norms on international assistance and cooperation. . . . Such norms would open new and promising avenues for handling the environmental consequences of war."

In April of 2015, a joint EU-UN-World Bank needs assessment estimated that a two-year effort to assess and respond to the most urgent environmental damage caused by the ongoing Ukraine conflict would cost $30 million. The eventual cost of restoration was impossible to calculate.

The broader context for the eventual remediation of the environmental damage should include the radical modernization of the region's notoriously unsustainable industry, which has for years presented direct and grave risks for its environment and people. In this way, this highly unwelcome conflict may, in the end, offer a rare and welcome opportunity to eventually "green" the black and brown coalfields of Donbas.

Syria: Cities Reduced to Toxic Rubble

Wim Zwijnenburg and Kristine te Pas, Pax for Peace (2015)

Prior to the 2011 uprising, the population of Syria was 21.5 million, of whom approximately one-third lived in rural areas. Deserts make up most of eastern Syria, with roughly one-third of the country fertile land. Around 55 percent of the country comprises natural pastures, steppe, desert, and mountainous areas. Natural resources include petroleum, phosphates, chrome and manganese ores, asphalt, iron ore, rock salt, marble, gypsum, and hydropower.

In spite of some animal and plant species having gone extinct, preconflict assessments in Syria reported a rich biodiversity. The country had twenty-six protected areas where mammals could not be hunted. (Although there are no designated natural parks, there were plans to establish national protected areas to ensure the conservation of fragile ecosystems.)

Between 2006 and 2010, Syria went through five successive years of drought. This had a serious impact on the agricultural sector. Around 1.3

million people were affected, and an estimated 800,000 famers and herders lost almost all their livestock. An estimated 300,000 farmer families migrated to the cities. Many resettled in informal urban housing areas, leaving them more vulnerable to environmental hazards caused by poor air quality and contaminated drinking water.

Targeting the Urban Landscape

Wars leave a toxic footprint in their wake. Considering the scale and duration of the ongoing Syrian conflict, it is beyond doubt that military targeting decisions, the widespread damage to populated areas, the vast quantity of weapons and munitions expended, and the breakdown of environmental services and governance have caused grave environmental damage, with long-lasting effects that will constrain post-conflict recovery in Syria.

Since the outbreak of hostilities in 2011, Syrians have witnessed violence and destruction on an enormous scale. As of October 2015, the fighting in Syria had caused the deaths of more than 210,000 people, with more than four million forced to flee their homes.

Aleppo, one of the oldest continuously inhabited cities on Earth and a UNESCO World Heritage Site, has suffered severe damage. Of its three million inhabitants in 2011, 1.8 million have been displaced, both in and outside the city. Since the outbreak of the civil war, more than 52 percent of Aleppo's housing units have been destroyed or suffered partial damage. Of its 123 neighborhoods, 21 have been left uninhabitable, while 53 are only partially functioning.

Damage to properties has generated millions of tons of rubble, which can contain a number of hazardous materials—including asbestos, cement, heavy metals, domestic chemicals, and combustion products—that can have detrimental effects on the environment and public health.

The national Syrian Arab Army (SAA) and its allied militias have increasingly resorted to the use of heavy-caliber, explosive weapons and cluster munitions in populated areas, and have even used chemical weapons.

Mortar bombs, artillery shells, barrel bombs, aircraft bombs, and missiles are conventional weapons that detonate to affect an area with blast and fragmentation. Data indicate that, worldwide, approximately 90 percent of those killed and injured when explosive weapons are used in populated areas are civilians.

Rebel forces, using a wide variety of conventional—often-improvised—weapons, have also carried out serious abuses, including indiscriminate attacks on civilians. In its attempt to create and hold a caliphate state encompassing much of Syria and northern Iraq, the so-called Islamic State (IS) jihadist army has used a range of tactics considered contrary to international humanitarian law, including the indiscriminate shelling of besieged towns, mass executions, slavery, and the use of mustard gas.

Barrel Bombs and Rockets

Intense fighting in urban areas involves the use of a variety of small- and medium-caliber munitions, explosives from mortars, artillery rounds, bombs, rocket-propelled grenades, and surface-to-surface and air-to-surface missiles. Low-order detonations—i.e., not fully detonated bombs—can result in the leaking of explosives such as RDX, DNT, and TNT, contaminating soil, surface water, and groundwater.

Aside from conventionally manufactured weapons, the SAA have also used so-called "barrel bombs"—large oil drums, gas cylinders, and water tanks filled with high explosives such as RDX, TNT, and scrap metal for fragmentation effects. Barrel bombs (some weighing more than 1,500 kilograms, or 1.7 tons) tend to contain a large amount of explosive material. Due to the difficulty of delivering them accurately to a target, barrel bombs tend to cause significant collateral damage.

The use of improvised rockets has been documented since 2012. These rockets are made by attaching rocket motors to large bombs, which are then launched indiscriminately into urban areas. Large versions are nicknamed "elephant rockets" because of their size and the noise they make when launched. They are short-range, destructive, and inaccurate.

Turning Ancient Cities into "Conflict Rubble"

When buildings are hit by munitions or damaged through pressure waves resulting from explosions, building materials are pulverized, generating large amounts of dust. Pulverized building materials (PBMs) are typically a mixture of cement, metals, polychlorinated biphenyls (PCBs), silica, asbestos, and synthetic fibers. Exposure to these materials can have a serious impact on health, both during the conflict and during post-conflict disposal.

Concrete is generally made of Portland cement mixed with water and coarse aggregates. Portland cement is a mixture of oxides of calcium, aluminum, iron, silicon, and magnesium. It also may contain selenium, thallium, and other impurities, all of which can become environmental contaminants and a hazard to human health.

The few studies that have been undertaken have found the rubble formed during conflict to have a higher proportion of combustion products, such as dioxins and furans, than typical demolition rubble. Potentially toxic PBMs are generated by high temperatures resulting from explosions.

Years of fighting have generated vast quantities of rubble and debris from damaged buildings in many districts in Syria's cities. One December 2013 investigation reported that 1.2 million houses—or one-third of all homes in Syria—had been damaged or destroyed.

Before the war, Homs was Syria's third-largest city, with more than 800,000 inhabitants. Famous for its Old City, Homs was a tourist hotspot known for its multicultural mix of communities. A May 2014 UN-Habitat assessment concluded that 50 percent of the neighborhoods in Homs had been heavily damaged and 28 percent partially damaged.

Heavy fighting erupted in 2011 and continued until government forces reclaimed control of the city in 2014, after a two-year siege. Almost half a million inhabitants were displaced and the intense fighting has had an enormous impact on the city. UN-Habitat analyses and UNISAT satellite damage assessments reveal severe destruction in the city's residential areas. More than 54 percent of Homs' housing is uninhabitable, and 26 of its 36 neighborhoods are completely or partially non-functional.

In addition to large numbers of people killed and injured directly from explosive weapons use, still more are affected by the damage that explosive weapons do to essential infrastructure such as schools, hospitals, housing, and water and sanitation systems. Living under bombardment also causes severe psychological distress, which often continues to impact the lives of those affected even after they have fled the area or the conflict has ceased.

The risk of explosive remnants of war can remain for decades after a conflict has ended. Munitions residues spread out over urban or rural areas during heavy fighting could include heavy metals such as tungsten, lead, or even depleted uranium, as well as energetic materials that make up explosives, such as RDX, PBX, and TNT and highly toxic rocket propellants. The targeting of military storage facilities during prolonged military campaigns in and around Aleppo also may have generated localized chemical hazards as well as unexploded ordnance (UXO).

Prolonged heavy fighting can leave behind pockets of contamination including heavy metals from munitions and toxic munitions constituents. Civilians remaining in or returning to these areas may be at risk of mixed exposures to munitions residues and pulverized building materials.

Damage to Industrial Areas

Direct damage to industrial sites in Homs, Hama, Damascus, and Aleppo has been reported, with buildings destroyed, burned, or looted, while some factories have been taken over by armed groups.

Around 90 percent of the pharmaceutical industry's facilities were located in Aleppo, Homs, and rural Damascus. It is estimated that twenty-five pharmaceutical plants have been destroyed. Aleppo's critical infrastructure has all but collapsed, as has law and order. The al-Sheikh Najjar industrial zone, situated fifteen kilometers (nine miles) from Aleppo, was once on its way to becoming the biggest industrial zone in the Middle East. It was intended to host 6,000 companies at its establishment in 2004, and 1,250 were in operation when the conflict started in 2011.

The destruction of water, sewage, and electricity systems can have serious repercussions. The U.S. bombing of electrical power facilities in the 1991 Gulf War shut down Iraq's water purification and sewage treatment plants—leading to outbreaks of gastroenteritis, cholera, and typhoid—and is thought to have caused an estimated 100,000 civilian deaths and a doubling of the infant mortality rate.

Water supply networks have regularly been targeted in the Syrian conflict. Dams, water pipes, and waste treatment plants have been damaged or destroyed due to attacks and counterattacks, by both regime and rebel forces.

A pumping station in Al Khafsah in Aleppo stopped working on May 10, 2014, after a military attack. This incident caused panic in the city, when nearly three million people lost access to water. A similar incident was reported in Aleppo, on November 9, 2012, when aerial bombing wrecked a water pipeline. Residents of Hama and Homs lost their water supply for several weeks, increasing the risk of waterborne diseases. In Aleppo, damage to the sewage system, which resulted in the contamination of drinking water, posed a serious risk to the population's health, as the price of fuel has skyrocketed, limiting the ability to boil water.

Many power plants have been damaged. Loss of power has had a serious impact on the water distribution and sanitation systems, which were already inadequate prior to the conflict. Many pumping stations have been damaged and are unable to operate. The situation became particularly serious during the first half of 2014, when large areas of the Aleppo and Deir ez-Zor governorates were completely cut off from running water. According to the World Health Organization, the availability of safe water in July 2014 was one-third that of pre-crisis levels.

Damage to the Power Grid

Syria's electricity network has increasingly become a military target during the conflict. By early 2013, Syria's minister of electricity claimed that more than 30 power stations were inactive and at least 40 percent of the country's high-voltage power lines had been attacked.

In August 2014, a group armed with shells, mortars, and light weapons targeted a plant near Hama in an attempt to take control of it. As the most prominent power plant in the central area, it was considered a key target. The plant was severely damaged.

The same plant was attacked again in November 2014, with rockets and shells causing further damage. One shell hit a diesel oil tank, causing a fire that consumed an estimated 1,350,000 liters (357,000 gallons) of diesel. Other power plants affected by the conflict include a thermal power plant east of Aleppo, the Zeyzoun power plant between Idlib and Homs, and the al-Zara plant between Homs and Hama. The latter two were severely damaged.

Because transformers contain PCBs, conflict damage has the potential to cause the release of PCBs into the environment from damaged power stations, substations, and distribution stations.

Toxic Clouds: The Smog of War

An attack on a government munitions depot in August 2013 resulted in a major explosion that killed forty people. Widespread dispersal of a range of munitions constituents, heavy metals, and propellants is likely to present a long-term local environmental and public health concern.

Munitions, explosives, and other military materials contain a range of potentially hazardous elements and compounds. Common metal constituents in small and light weapons ammunition include lead, copper, mercury, antimony, and tungsten. Lead makes up 95 to 97 percent of the metallic components of military ammunitions and grenades.

Toxic materials prevalent in munitions include dinitrotoluene (DNT), trinitrotoluene (TNT), hexahydrotrinitrotriazine (RDX), and octahydrotetranitrotetrazocine (HMX). Other toxic substances often found in weapons include solid or liquid propellants for various types of rockets and missiles, such as hydrazine, nitroglycerin, nitroguanidine, nitrocellulose, and 2,4-dinitrotoluene. Various perchlorate formulations are employed in missile, rocket, and gun propellants.

Most explosive compounds are relatively persistent in the environment. TNT may be transformed by sunlight or microbial action into compounds more toxic than itself. RDX, HMX, and perchlorate appear to be common groundwater contaminants, while TNT is generally not. Whether solid, liquid, or vaporized, these substances have the potential to harm human health, depending on dose, duration, and route of exposure.

Conflicts involving state and non-state actors make determining strict liability and accountability for environmental damage and its humanitarian consequences challenging, but states and civic society must develop new systems of response and assistance to improve the protection of civilians and the environment upon which they depend.

Wars and Refugees

Tom H. Hastings (2017)

According to the United Nations, as of early 2017, armed conflicts had forcibly displaced 65.3 million people, with some fleeing to other parts of their own lands while others joined a tide of 21 million seeking safety in other countries. More than half of those fleeing conflict and persecution came from just three war-torn nations—Syria, Afghanistan, and Somalia. It is much harder to flee the horrific conflict in Yemen, where war has forced 3 million from their homes and 2.2 million children from their schools, and left 7.3 million undernourished in a country where Saudi aircraft routinely drop U.S.-supplied cluster bombs on civilian populations.

When the hail of bullets and rain of bombs begin, the flood of targeted humanity gathers and soon starts to run. Like so many other human-caused phenomena, refugee flight was not the real environmental problem when the Earth was a sea of wilderness dotted with islands of human disturbance. Now, however, the situation is reversed and the oceans of humanity cause serious damage when whipped up by the storms of war.

This has not been a mere side effect of armed human conflict; the wall of displaced humanity rolls over a countryside not as a ripple effect but as a tidal wave of great and destructive force. Like any flood, the fresh becomes the dirty, the clean becomes the unhealthy, and entire habitats are swallowed up in diseased destruction.

Beautiful little children and caring women—the main channel of most rivers of war refugees—become part of cholera-producing, typhus-carrying, eco-trampling, resource-gobbling waves of misery. Often, they precipitate even more conflict, into a kind of mutually reinforcing dynamic of tragedy. Approximately 34,000 people—more than half of whom are under age 18—are driven from their homes daily.

From Burma to Rwanda, from Somalia to Afghanistan, the tremendous numbers add up to half a percent of humanity, nearly one in every 200 of us on Earth. More than 90 percent of the armed conflict since World War II—despite the high-profile interstate shooting wars—has been internal. While bombs, bullets, and land mines have taken a tremendous toll, the direct cutoff of food and potable water has taken an even greater toll in those conflict areas. In the conflict zones, more than 100 million people are chronically malnourished.

The problem of war refugees became worse, rather than better, in the final quarter of the twentieth century and then dramatically worsened again in the new millennium. In the early 1960s, the number of international refugees was estimated at a bit more than one million. The United Nations High Commissioner for Refugees (UNHCR) reported 2.8 million international refugees in 1976. That figure swelled to nearly 19 million by 1993.

In 1997, an astonishing 63 percent of Liberians were refugees. Forty-five percent of Rwandans and Bosnians were refugees, as were 20 percent of all Afghanis. To imagine the impact of the returning Jordanians and Palestinians—evicted from Kuwait following the war—visualize the crisis if the United States were forced to embrace a sudden increase of 30 million in one year—the proportional equivalent to Jordan's 12 percent growth due to immigration of war refugees.

In 2015, the UNHCR reported that wars in hotspots like Sudan, Yemen, and Syria had helped drive the number of desperate refugees to a global record of 59.5 million. In 2017, 4.9 million Syrians, 2.7 million Afghans, and 1.1 million Somalis fled over their borders, and the numbers just increase as the world fails to rein in destructive conflict.

One of the most opportunistic and cynical treatments of refugees is to turn them into proxy troops, as the United States did, first to Cubans fleeing Castro and, two decades later, Somocistas running from Sandinistas in Nicaragua. Civilians in flight from civil war in the early 1970s were armed by the Indian government and pitted against Pakistan just as Afghanis seeking refuge were given guns and propelled against the Soviets. The Russians did the same to Chechens fleeing from Grozny in 1999. Palestinians have historically been trained and armed by Arab states to fight Israel, and Kurds have been used with little regard by both Iran and Turkey, against each other.

Compounding the problems of refugee movement is the ecological and economic lack of skill when demobilization succeeds the peace accord. The dispossessed know how to make their living with their weapons, and everyone is telling them to stop carrying the only tools they know how to use. Making war has taught them to abuse the land, not care for it. Taught to believe that the greatest goal is destruction, now they are being told to stoop to encourage delicate life.

"Most of them did not have any profession. . . . And now—in a disarmed situation . . . the only thing that they know how to do is to kill." So said Graca Machel, Mozambican children's rights spokesperson and widow of Nelson Mandela.

We need only look at examples of attempted or temporary demobilization—Afghanistan, Angola, Cambodia—to see the hardball facts: declaring peace or a cease-fire is not enough. The war may be over, cantonment may be called complete, but the land is still suffering from buried mines and the other dangers. While former army or guerrilla soldiers are reluctantly learning to make an honest living (many still eager for the swashbuckling days of parading with weaponry; of feeling powerful in a frightening time), they often do a poor job at tedious farm and factory labor. Soil exhaustion, erosion, and the overuse of chemicals is part of the mark of a former military man behind the tractor, spray-rig, or hoe. As the Earth's ecology continues to unravel, so does every peace accord signed in hope.

Between 2008 and 2015, more than 22 million people were displaced by floods, hurricanes, wildfires, and rising seas. With anthropogenic climate chaos now upon us, scores of millions more stand to lose their homes, lands, cities, even entire island nations in the coming twenty to thirty years.

When we solve the problem of war, all the rest will be far less difficult. The root problem lies in conflict management. Instead of turning to generals, security studies experts, and politicians whose campaigns are funded by war contractors, it is long past time to turn to activist leaders in civil society, to scholars of strategic nonviolence, and to elected officials untainted by war profits. With a serious public discourse on alternative methods of security—focused on constructive creative conflict management methods—we can begin to staunch the flow of blood, refugees, and environmental devastation.

Civilian Victims of Killer Drones

Medea Benjamin, CODEPINK (2013)

I met Roya on my first day while visiting the Pakistan-Afghan border, on a dusty road in Peshawar. It was just weeks after the 2002 U.S. invasion of Afghanistan, and I was traveling as a representative of the human rights group I cofounded called Global Exchange. A young girl approached me, her head cocked to one side, her hand outstretched, begging for money.

With the help of an interpreter, I learned her story. Roya was thirteen years old, the same age as my youngest daughter. But her life could not have been in starker contrast to that of my San Francisco high schooler and her girlfriends. Roya never had time for sports, or for school. She was born into a poor family living on the outskirts of Kabul. Her father was a street vendor; her mother raised five children and baked sweets for him to sell.

One day, while her father was out selling candies, Roya and her two sisters were trudging home carrying buckets of water. Suddenly, they heard a terrifying whir and then there was an explosion. Something terrible had dropped from the sky, tearing their house apart and sending the body parts of their mother and two brothers flying through the air.

The Americans must have thought Roya's home was part of a nearby Taliban housing compound. In the cold vernacular of military-speak, her family had become "collateral damage" in America's war on terror.

When Roya's father came home, he carefully collected all the bits and pieces of his pulverized family that he could find, buried them immediately according to Islamic tradition, and then sank into a severe state of shock.

Roya became the head of her household. She bundled up her surviving sisters, grabbed her father, and fled. With no money or provisions, they trekked through the Hindu Kush, across the Khyber Pass, and into Pakistan.

In Peshawar, the family barely survived on the one dollar a day the girls made from begging. Roya took me to their one-room adobe hut to meet her father. A tall, strong man with the calloused hands of a hard worker, he no longer works. He doesn't even walk or talk. He just sits and stares into space. "Once in a while he smiles," Roya whispered.

Inside Afghanistan, I saw more lives destroyed by U.S. bombs. Some bombs hit the right target but caused horrific collateral damage. Some bombs hit the wrong target because of human error, machine malfunction, or faulty information.

In one village, the Americans thought a wedding party was a Taliban gathering. One minute, forty-three relatives were joyously celebrating; the next minute, their appendages were hanging off the limbs of trees.

Forty villagers were killed in another small town in the middle of the night. Their crime? They lived near the caves of Tora Bora, where Osama bin Laden was presumed to be hiding. The U.S. news media reported the dead as Taliban militants. But the woman I met—who had just lost her husband and four children, as well as both her legs—had never heard of al-Qaeda, America, or George Bush. Bleeding profusely, she was praying that she would die. Surviving as a crippled widow with no income and no family was too much to bear.

The Navy's Sonic War on Whales

Pierce Brosnan, Natural Resources Defense Council (2014)

Whales and other marine mammals rely on their hearing for life's most basic functions, such as orientation and communication. Sound is how they find food, find friends, find a mate, and find their way through the world every day. So when a sound thousands of times more powerful than a jet engine fills their ears, the results can be devastating—and even deadly.

This is the reality that whales and other marine mammals face because of human-caused noise in the ocean, whether it's the sound of air-guns used in oil exploration or subs and ships emitting sonar. Man-made sound waves can drown out the noises that marine mammals rely on for their very survival, causing serious injury and even death.

If you've ever seen a submarine movie, you probably came away with a basic understanding of how sonar works. Active sonar systems produce intense sound waves that sweep the ocean like a floodlight, revealing objects in their path.

Some systems operate at more than 235 decibels, producing sound waves that can travel across tens or even hundreds of miles of ocean. During testing off the California coast, noise from the Navy's main low-frequency sonar system was detected across the breadth of the northern Pacific Ocean.

By the Navy's own estimates, even 300 miles from the source, these sonic waves can retain an intensity of 140 decibels—a hundred times more intense than the level known to alter the behavior of large whales.

The Navy's most widely used sonar systems operate in the mid-frequency range. Evidence of the danger caused by these systems surfaced dramatically in 2000, when whales of four different species stranded themselves on beaches in the Bahamas. Although the Navy initially denied responsibility, the government's investigation established that mid-frequency sonar caused the strandings.

After the incident, the area's population of Cuvier's beaked whales nearly disappeared, leading researchers to conclude that they either abandoned their habitat or died at sea. Similar mass strandings have occurred in the Canary Islands, Greece, Madeira Island, the U.S. Virgin Islands, Hawaii, and other sites around the globe.

Many of these beached whales have suffered physical trauma, including bleeding around the brain, ears, and other tissues and large bubbles in their organs. These symptoms are akin to a severe case of "the bends"—the illness that can kill scuba divers who surface quickly from deep water. Scientists believe that the mid-frequency sonar blasts may drive certain whales to change their dive patterns in ways their bodies cannot handle, causing debilitating and even fatal injuries.

Stranded whales are only the most visible symptom of a problem affecting much larger numbers of marine life.

In 2014, the Navy announced a five-year sonar-training plan that could threaten entire populations of marine wildlife off the East Coast, Southern California, Hawaii, and the Gulf Coast. The Navy estimates the increased sonar training will significantly harm marine mammals more than 10 million times off the U.S. coast alone. Navy ships will also conduct torpedo tests, bombing exercises, and underwater explosions—some 1.1 million events overall; an average of one detonation every two minutes for five years.

As a direct result, nearly 1,000 marine mammals could die. There could be more than 13,000 serious injuries—including permanent hearing loss and lung damage. And that's according to the Navy's own numbers.

Naval sonar has been shown to disrupt feeding and other vital behavior and to cause a wide range of species to panic and flee. Scientists are concerned about the cumulative effect of all of these impacts on marine animals.

A History of Whale Strandings

May–June 2015—Six large whales (humpbacks and grays) wash up on California's beaches from Santa Cruz to San Francisco over a two-month period. All show signs of muscle hemorrhaging and bleeding of the brain—possible signs of harmful sonar exposure. The U.S. Navy declined to provide data on sonar drills along the coast.

April 2014—Seven Cuvier's beaked whales began stranding along the southern coast of Crete as the U.S., Greek, and other naval vessels engage in Operation Noble Dina war games offshore.

July 2011—Seventy pilot whales are driven onto Scottish shores after Britain's Royal Navy explode four undersea bombs at Cape Wrath, Europe's largest live-bombing range—nineteen died.

May–June 2008—A mass stranding of approximately 100 melon-headed whales off northwest Madagascar was triggered by a multi-beam echo-sounder system operated by a survey vessel contracted by ExxonMobil.

January 2006—At least four beaked whales strand in the Gulf of Almeria, Spain, while sonar exercises take place offshore.

January 2005—At least thirty-four whales of three species strand along the Outer Banks of North Carolina as Navy sonar training goes on offshore.

July 2004—Four beaked whales strand during naval exercises near the Canary Islands.

July 2004—Approximately 200 melon-headed whales crowd into the shallow waters of Hanalei Bay in Hawaii as a large Navy sonar exercise takes place nearby. Rescuers succeed in directing all but one of the whales back out to sea.

June 2004—As many as six beaked whales strand during a Navy sonar-training exercise off Alaska.

May 2003—As many as eleven harbor porpoises beach along the shores of the Haro Strait, Washington State, as the *USS Shoup* tests its mid-frequency sonar system.

September 2002—At least fourteen beaked whales from three different species strand in the Canary Islands during an anti-submarine warfare exercise in the area. Four additional beaked whales strand over the next several days.

May 2000—Three beaked whales strand on the beaches of Portugal's Madeira Island during NATO naval exercises.

March 2000—Seventeen cetaceans, including fourteen beaked whales, beach themselves and die during Navy sonar exercises in the waters off Bermuda.

October 1999—Four beaked whales strand in the U.S. Virgin Islands during Navy maneuvers offshore.

October 1997—At least nine Cuvier's beaked whales strand in the Ionian Sea, with military activity reported in the area.

May 1996—Twelve Cuvier's beaked whales strand on the west coast of Greece as NATO ships sweep the area with low- and mid-frequency active sonar.

October 1989—At least twenty whales of three species strand during naval exercises near the Canary Islands.

December 1991—Two Cuvier's beaked whales strand during naval exercises near the Canary Islands.

In 2013, the Navy revealed that live munitions training—scheduled to take place from 2014 to 2019—was expected to kill 186 whales and dolphins off the East Coast and 155 off Hawaii and Southern California. There could be 11,267 serious injuries and 1.89 million minor injuries (such as temporary hearing loss) off the East Coast alone, while exercises in the waters off Hawaii and Southern California could cause 2,039 serious injuries and 1.86 million temporary injuries.

On September 15, 2015, the NRDC and other environmental groups announced the Navy had agreed to limit its use of sonar and explosives to avoid harming whales, dolphins, and other marine mammals. The agreement limits or bans the use of mid-frequency active sonar and explosives in specified areas around the Hawaiian Islands and Southern California.

A FIELD GUIDE TO MILITARISM

When you live in the United States, with the roar of the free market, the roar of this huge military power, the roar of being at the heart of empire, it's hard to hear the whispering of the rest of the world.

—Arundhati Roy, Booker Prize–
winning author and activist

The Militarization of Native Lands

Winona LaDuke, Honor the Earth (2013)

The Department of Defense states that, as part of carrying out its mission to defend America, "certain activities—such as weapons testing, practice bombing and field maneuvers—may have had effects on tribal environmental health and safety as well as tribal economic, social and cultural welfare." That would be an understatement.

Blowing things up (not cleaning up after itself) is the military's strong suit. Only a tiny fraction of the federal defense budget is spent on resolving the Pentagon's local and global environmental impacts.

According to a 2004 Associated Press story, "removing unexploded munitions and hazardous waste found so far on 15 million acres of shut-down U.S. military ranges could take more than 300 years, according to congressional auditors." This cost, originally estimated at $35 billion, was soon climbing rapidly. In one report, the military identified 17,482 potentially contaminated sites at 1,855 installations. Rather than clean up the toxins, the Department of Defense simply limited its liability for much of the contamination.

The military is one of the largest landholders in the United States, with some 30 million acres of land under its control. Much of this land base was annexed or otherwise stolen from Native peoples. The two states with the most federal landholdings are Nevada with 84.5 percent of the state and Alaska with 69.1 percent of the state being held by the federal government. These represent takings under the 1863 Ruby Valley Treaty with the Shoshone and the Alaskan Native Claims Settlement Act of 1971.

In 1916, the U.S. Army owned approximately 1.5 million acres. Land ownership grew by 33 percent in the course of the World War I mobilization. As of 1940, the Army owned approximately two million acres. The scale of World War II mobilization was unprecedented: the Army (including the Army Air Forces) acquired eight million additional acres, thereby quintupling the Pentagon's land ownership. Military agencies were given nearly unlimited spending authority to acquire land from private owners and had the option of foreclosing on lands it deemed necessary for national security.

Most of the new acreage cost the Army practically nothing. More than six million acres—more than three-quarters of the land it acquired—came from the public domain.

What is clear is that "public domain" is often really Native land. The Army took some 10,000 acres from the Cheyenne and Arapaho people for Fort Reno in 1881 and, until 1993, had use of Zuni and Navajo lands near Fort Wingate.

In fact, much of what is today U.S. military land was, at some time, taken from Native peoples, sometimes at gunpoint, sometimes in the wake of massacres or forced marches, sometimes through starvation, and sometimes through pen and paper—broken treaties, acts of Congress or state legislatures, or by presidential authority by the "Great White Father" himself.

Native Lands as Dumpsites

Until 1940, the Army's Chemical Warfare Service did all its testing at the Edgewood Arsenal in New Jersey and the Aberdeen Proving Ground in Maryland. Eventually these facilities became overcrowded, too close to growing population centers, and too small for large-scale testing of toxic agents. The Army's answer: move their chemical warfare testing to the heart of Goshute Territory.

About 70 miles southwest of Salt Lake City, Utah, a small community of Goshutes lives on an 18,600-acre reservation. For the past forty years, the federal government has created, tested, and dumped toxic military wastes all around them. Less than ten miles southwest of the reservation is the Dugway Proving Grounds, where the government conducts tests of chemical and biological weapons.

In 1968, chemical agents escaped from Dugway and killed more than 6,000 sheep and other animals while the Army was conducting open-air nerve gas tests that included delivery by artillery shells and fighter jet. More than 1,600 of those animals were buried on the reservation, leaving a toxic legacy in the ground.

At least 1,174 other tests of chemical agents at Dugway spread nearly a half-million pounds of nerve agent to the winds. There were 328 open-air germ warfare tests, 74 radiological "dirty bomb" tests, and the equivalent of eight intentional meltdowns of small nuclear reactors.

The Desert Chemical Depot, located 15 miles east of the reservation, stores more than 40 percent of the nation's chemical weapons stockpile and is responsible for the incineration of many chemical munitions and nerve agents.

The Skull Valley Goshutes were never consulted about the placement of any of these facilities, nor have they ever been compensated for the immense threats to their environment and health that these sites pose. Their story is a microcosm of the impact of the military on Native people.

Pueblo Land: Wounded by Nuclear Bombs

New Mexico's infamous Los Alamos National Laboratory has had more than its share of fires and toxic leaks. The laboratory was built on more than forty-three square miles of land taken from the Santa Clara and San Ildefonso Pueblos.

The multitude of nuclear weapons research facilities has left a series of nuclear dumps, including some 17.5 million square feet of hazardous and nuclear waste at 24 officially designated storage areas. These sites include plutonium and tritium plants used to build parts of nuclear warheads. Thus far, the United States has expended $7 trillion on nuclear warheads, a good sum of this allocated to Los Alamos. Of the laboratory's $2.1 billion FY 2014 budget, some 73 percent was slated for nuclear weapons work. No long-term plan for waste disposal or reduction exists.

Alaska: Occupied Native Territory

Alaska's Natives have felt some of the most widespread and deepest impacts of the modern military. Alaska has over 200 Native villages and communities and almost as many military sites. The military holds over 1.7 million acres of Alaska, much of this within the traditional territories of Indigenous communities. Then there is the military occupation of airspace over Native territories where low-flying planes and sonic booms can have a disruptive impact on caribou herds.

Seven hundred active and abandoned military sites account for at least 1,900 toxic hotspots. Five out of Alaska's seven Superfund sites are a result of military contamination. The 700 formerly used defense sites in Alaska tell a history of the Cold War and every war since. The levels of radioactive and persistent organic pollutants remaining in the environment continue to put at risk people who are dependent upon the land for their survival.

Alaskan Native lands were occupied for military reasons but, in many cases, were annexed for their oil and natural resources under the Alaskan Native Claims Settlement Act of 1971. The military remains dominant on the land, joined by multinational oil, mining, and logging companies.

Western Shoshone: Nuclear Tests and Atomic Waste

In 1951, the Atomic Energy Commission created its Nevada Test Site (NTS) within Western Shoshone Territory as a "proving ground" for nuclear weapons. Between 1951 and 1992, the United States and Great

Britain exploded 1,054 nuclear devices both above and below ground at the NTS. Radiation from these experiments was fully measured for only 111 of the tests, about 10 percent of the total.

In 1997, the National Cancer Institute (NCI) released a study of radiation exposure from aboveground nuclear tests that revealed some 160 million people suffered significant radiation exposure from the tests—on average 200 times more than the amount indicated by the government. The NCI estimated that as many as 75,000 cases of thyroid cancer may have been caused by atmospheric testing.

Hawaii: Native Lands Seized by Military

The military controls a large percentage of Hawaii, including some 25 percent of Oahu, valuable "submerged lands" (i.e., estuaries and bays) and, until relatively recently, the island of Kaho'olawe. The military reigns over more than 200,000 acres of Hawaii, with over 100 military installations and at least 150,000 personnel.

Live ammunition occasionally washes up on local beaches. Malu Aina, a military watchdog group from Hawaii, reports: "Live military ordnance in large quantities has been found off Hapuna Beach and in Hilo Bay. Additional ordnance, including grenades, artillery shells, rockets, mortars, armor-piercing ordnance, bazooka rounds, napalm bombs, and hedgehog missiles have been found at Hilo airport, in Waimea town, Waikoloa Village . . . and on residential and school grounds. At least nine people have been killed or injured by exploding ordnance."

Since the end of World War II, Hawaii has been the center of the U.S. military's Pacific Command. PACOM serves as an outpost for Pacific expansionism, along with Guam, the Marshall Islands, Samoa, and the Philippines. It is the center of U.S. military activities over more than half the Earth—from the west coast of United States to Africa's east coast, from the Arctic to Antarctica, covering 70 percent of the world's oceans.

The island of Kaho'olawe was the only national historic site to be used as a bombing range. After years of litigation and negotiations, Congress placed a moratorium on the bombing, but after $400 million in cleanup money, much remains to be completed. The military determined that there are more than 236 former military sites in Hawaii at 46 separate installations, all of which were contaminated.

There is no way to avoid an observation: it is as a result of our nation's history of colonialism, the Doctrine of Manifest Destiny, and the expansion of military interests to support American imperialism that Indian Country communities are located adjacent to more than our fair share of these toxic military sites. That is because many of today's military bases are the legacy of old U.S. Calvary forts, places where the Army built strongholds to support their invasions of Indian Country, places that were used to subjugate and imprison Native people.

The Yakima Nation: A Reservation for Nuclear Wastes

The Hanford Nuclear Reservation—the largest nuclear waste dump in the Western Hemisphere—is located entirely within the treaty boundary of the Yakama Nation. As Hanford Watch activist Lynne Porter reports: "Hanford covers 560 square miles of high brush lands in eastern Washington, along 51 miles of the Columbia River. From 1943 to 1988, Hanford produced plutonium for nuclear weapons, using a line of nuclear reactors along the river. About 53 million gallons of high-level radioactive and chemical waste are stored in 177 underground tanks (the size of three-story buildings) buried in Hanford's central area, about 12 miles from the river. Over the years, 70 of the tanks have leaked about 1 million gallons of waste into the soil."

Umatilla Land: Chemical and Ammunition Depots

For 12,000 years, the Umatilla, Wasco, Cayuse, and Walla Walla have lived in the Columbia Plateau on lands and waters that sustained their people. In 1850, however, Congress passed the Oregon Donation Land Act, exclusively for the benefit of white "settlers" and "pioneers," and thereby created the largest land giveaway in the history of the nation—at the sole expense of Native people, who paid with their lives by the thousands. Relentlessly harassed and murdered by both U.S. Army troops and bloodthirsty white vigilantes, surviving remnants were forced to sign treaties under egregious terms in 1855 and 1856.

In 1940, the Army selected a 16,000-acre parcel from within this territory to become the Umatilla Ordnance Depot. Beginning in 1941, some 7,000 workers were hired and $35 million was spent to create a complex of military storehouses, housing, and ammunition. The munitions that were stored there were used in the Korean conflict, Vietnam, Grenada, Panama, Operation Desert Shield, and Operation Desert Storm. In 1962, chemical nerve agents VX and GM and the mustard blister agent HD were sent there. The Umatilla Chemical Agent Disposal Facility covers more than 19,000 acres of Confederated Tribes of Umatilla land.

The Great Lakota Nation and the Badlands Bombing Range

It is said that if the Great Sioux Nation were in control of its 1851 treaty area, it would be the third-greatest nuclear weapons power on the face of the Earth. This is due to the vast number of Air Force, NORAD, and other bases in the Lakota territories now called Nebraska, North and South Dakota, Wyoming, and Montana. Of particular concern to the Lakota Nation, however, is the gunnery range on the Pine Ridge reservation, which is part of a more recent land seizure. As Sam Featherman explains in the book *The Treadmill of Destruction*: "The U.S. government seized 342,000 acres of the Pine Ridge reservation in South Dakota for a bombing range to train World War II pilots. The land seizure forced fifteen Oglala Sioux families to sell their farms and ranches for $.03 an acre." The gunnery range (aka the Badlands Bombing Range) continues to be a source of concern for the Oglala Sioux Tribe, as both live and spent ordnance are found throughout the area.

The Ho-Chunk Nation and Badger Munitions

The Badger Army Ammunitions plant near Baraboo, Wisconsin, was established by the military in 1942 as a Class II military propellant manufacturing installation. The plant produced propellants for small arms, rockets, and a host of larger weapons. Ammunition production ceased in 1975, but the plant occupied 7,400 acres of land, some contaminated with a host of toxic chemicals including chloroform, carbon tetrachloride, trichloroethene, and dinitrotoluene. Over the past two decades, the

Ho-Chunk Nation has been seeking return of the lands associated with the former Badger facility. With this land, the Ho-Chunk plan to create an organic bison operation to feed the tribal population.

Wars Are No Longer Fought on Battlefields

David Swanson, World Beyond War (2010)

We talk of sending soldiers off to fight on battlefields. The word "battlefield" appears in news stories about our wars. And the term conveys to many of us a location in which soldiers fight other soldiers. We don't think of certain things being found in a battlefield. We don't imagine whole families, or picnics, or wedding parties, for example, as being found on a battlefield—or grocery stores or churches. We don't picture schools or playgrounds or grandparents in the middle of an active battlefield. We visualize something similar to Gettysburg or World War I France: a field with a battle on it. Maybe it's in the jungle or the mountains or the desert of some distant land we're "defending," but it's some sort of a field with a battle on it. What else could a battlefield be?

At first glance, our battlefields do not appear to be where we live and work and play as civilians, as long as "we" is understood to mean Americans. Wars don't happen in the United States. But for the people living in the countries where our wars have been fought since, and including World War II, the "battlefield" has quite clearly included and continues to include their hometowns and neighborhoods.

While the Battles of Bull Run or Manassas were fought in a field near Manassas, Virginia, the Battles of Fallujah were fought in the city of Fallujah, Iraq. When Vietnam was a battlefield, *all* of it was a battlefield, or what the U.S. Army now calls "the battlespace." When our drones shoot missiles into Pakistan, the suspected terror plotters we're murdering are not positioned in a designated field; they're in houses, along with all of the other people we "accidentally" kill as part of the bargain. (And at least some of those people's friends will indeed begin plotting terrorism, which is great news for the manufacturers of drones.)

Pilots speak of being on the battlefield when they have been great distances *above* anything resembling a field or even an apartment building. Sailors speak of being on the battlefield when they haven't set foot on dry land. But the new battlefield also encompasses anywhere U.S. forces might conceivably be employed, which is where your house comes in. If the president declares you an "enemy combatant," you will not only live on the battlefield—you will be the enemy, whether you want to be or not.

When U.S. forces kidnap people on the street in Milano or in an airport in New York and send them off to be tortured in secret prisons, or when our military pays a reward to someone in Afghanistan for handing over their rivals and falsely accusing them of terrorism, and we ship the victims off to be imprisoned indefinitely in Guantanamo or right there in Bagram, all of those activities are said to take place on a battlefield. Anywhere someone might be accused of terrorism and kidnapped or murdered is the battlefield. No discussion of releasing innocent people from Guantanamo would be complete without expression of the fear that they might "return to the battlefield," meaning that they might engage in anti-U.S. violence, whether they had ever done so before or not.

When an Italian court convicts CIA agents in absentia for kidnapping a man in Italy in order to torture him, the court is staking the claim that Italian streets are not located in a U.S. battlefield. When the United States fails to hand over the convicts, it is restoring the battlefield to where it now exists: in each and every corner of the galaxy. Traditionally, killing people has been deemed legal in war but illegal outside of it. Apart from the fact that our wars are themselves illegal, should it be permissible to expand them to include an isolated assassination in Yemen? What about a massive bombing campaign with unmanned drones in Pakistan? Why should the smaller expansion of an isolated murder be less acceptable than the larger expansion that kills more people?

How the White House Can Wage War Inside the United States

And if the battlefield is everywhere, it is in the United States as well. The Obama administration in 2010 announced its right to assassinate Americans, presuming to already possess by common understanding the right to assassinate non-Americans. But it claimed the power to kill

Americans only outside the United States. Yet, active military troops are stationed within the United States and assigned to fight here if so ordered. If the "war on terror" makes it wartime, and if the "war on terror" lasts for generations, as some of its proponents desire, then there really are no limits.

Wars have always had a tendency to eliminate hard-won rights. In the United States, this tradition includes President John Adams's Alien and Sedition Acts of 1798, Abraham Lincoln's suspensions of habeas corpus, Woodrow Wilson's Espionage Act and Sedition Act, Franklin Roosevelt's rounding up of Japanese Americans, the madness of McCarthyism, and the many developments of the Bush-Obama era that really took off with the first passage of the PATRIOT Act.

On July 25, 2008, the House Judiciary Committee agreed to hold a hearing on the impeachment of George W. Bush. The hearing was just a stunt. But the testimony was deadly serious and included a statement from former Justice Department official Bruce Fein from which this is excerpted:

> After 9/11, the executive branch declared—with the endorsement or acquiescence of Congress and the American people—a state of permanent warfare with international terrorism [and that] . . . the entire world, including all of the United States, is an active battlefield where military force and military law may be employed at the discretion of the executive branch.
>
> For instance, the executive branch claims authority to employ the military for aerial bombardment of cities in the United States if it believes that Al Qaeda sleeper cells are nesting there and are hidden among civilians. . . .
>
> The executive branch has directed United States forces to kill or kidnap persons it suspects have allegiance to Al Qaeda in foreign lands, for instance Italy, Macedonia, or Yemen, but it has plucked only one United States resident, Ali Saleh Kahlah al-Marri, from his home for indefinite detention as a suspected enemy combatant. But if the executive branch's constitutional justification for its modest actions is not rebuked through impeachment or otherwise, a precedent of executive power will have been

established that will lie around like a loaded weapon ready for use by any incumbent who claims an urgent need.

President Obama maintained and expanded upon the powers established by George W. Bush. War was now officially everywhere and eternal, thereby allowing presidents even greater powers, which they could use in the waging of even more wars, from which yet more powers could derive, and so forth to Armageddon, unless something breaks the cycle.

The Battlefield Is Nowhere and Death Is Everywhere

The battlefield may be all around us, but the wars are still concentrated in particular places. Even in those particular locations—such as Iraq and Afghanistan—the wars lack the two basic features of a traditional battlefield—the field itself and a recognizable enemy. In a foreign occupation, the enemy looks just like the supposed beneficiaries of the "humanitarian" war. The only people recognizable for who they are in the war are the foreign occupiers.

Wars are not waged against armies. Nor are they waged against demonized dictators. They are waged against people. Remember the U.S. soldier who shot a woman who had apparently been bringing a bag of food to the U.S. troops? She would have looked just the same if she had been bringing a bomb. How was the soldier supposed to tell the difference? What was he supposed to do?

The answer, of course, is that he was supposed to not be there. The occupation battlefield is full of enemies who look exactly like, but sometimes are not, women bringing groceries. It is a lie to call such a place a "battlefield." One way to make this clear is to note that a majority of those killed in wars are civilians. A better term is probably "non-participants." Some civilians participate in wars. And those who resist a foreign occupation violently are not necessarily military. Nor is there any clear moral or legal justification for killing those fighting a truly defensive war any more than there is for killing the non-participants.

The "good war," World War II, is still the deadliest of all time, with military deaths estimated at 20 to 25 million (including 5 million deaths of prisoners in captivity), and civilian deaths estimated at 40 to 52 million (including 13 to 20 million from war-related disease and famine). The United States suffered a relatively small portion of these deaths—an estimated 417,000 military and 1,700 civilian—small in relation to the suffering of some of the other countries.

The War on Korea saw the deaths of an estimated 500,000 North Korean troops; 400,000 Chinese troops; 245,000 to 415,000 South Korean troops; 37,000 U.S. troops; and an estimated 2 million Korean civilians.

The War on Vietnam may have killed 4 million civilians or more, plus 1.1 million North Vietnamese troops, 40,000 South Vietnamese troops, and 58,000 U.S. forces.

In the decades following the destruction of Vietnam, the United States killed a lot of people in a lot of wars, but relatively few U.S. soldiers died. The Gulf War saw 382 U.S. deaths. The 1965–66 invasion of the Dominican Republic didn't cost a single U.S. life. Grenada in 1983 cost nineteen. Panama in 1989 saw forty Americans die. Bosnia-Herzegovina and Kosovo saw a total of thirty-two U.S. war deaths. Wars had become exercises that killed very few Americans in comparison to the large numbers of non-U.S. non-participants dying.

The wars on Iraq and Afghanistan similarly saw the other sides do almost all of the dying. Americans hear through their media that over 4,000 U.S. soldiers have died in Iraq, but rarely do they encounter any report on the deaths of Iraqis. Fortunately, some serious studies have been done of Iraqi deaths caused by the invasion and occupation that began in March 2003.

A study by *The Lancet,* based on surveys of Iraqi households through the end of June 2006, concluded that there had been 654,965 excess violent and nonviolent deaths. This included deaths resulting from increased lawlessness, degraded infrastructure, and poor healthcare. Most of the deaths (601,027) were estimated to be due to violence. The causes of violent deaths were gunshot (56 percent), car bomb (13 percent), other explosion/ordnance (14 percent), air strike (13 percent), accident (2 percent), and unknown (2 percent). Extrapolating from the *Lancet* report, Just Foreign Policy, a Washington-based organization, calculated that Iraqi deaths through 2011 totaled 1,366,350.

For all of these wars, one can add a much larger casualty figure for the wounded than those I've cited for the dead. It is also safe to assume in each case a much larger number for those traumatized, orphaned, made homeless, or exiled. The Iraqi refugee crisis involves millions. Beyond that, these statistics do not capture the degraded quality of life in war zones—reduced life expectancy, increased birth defects, the rapid spread of cancers, and the horror of unexploded bombs.

If the numbers above are correct, World War II killed 67 percent civilians; the War on Korea, 61 percent civilians; the War on Vietnam, 77 percent civilians; the War on Iraq, 99.7 percent Iraqis (whether or not civilians); and the Drone War on Pakistan, 98 percent civilians.

On March 16, 2003, a young American woman named Rachel Corrie stood in front of a Palestinian home in the Gaza strip, hoping to protect it from demolition by the Israeli military, which claimed to be destroying guerrilla hideouts. She faced a Caterpillar D9-R bulldozer, and it crushed her to death. Defending against her family's civil suit in court in September 2010, an Israeli military training unit leader explained: "During war there are no civilians."

Deaths at Checkpoints

In 2007, the U.S. military admitted to having killed 429 civilians at Iraqi checkpoints. In an occupied country, the occupier's vehicles must keep moving or those inside might be killed. The vehicles belonging to the occupied, however, must stop to prevent their being killed. War on Iraq veteran Matt Howard remembers:

> An American life is always worth more than an Iraqi life. Right now, if you're in a convoy in Iraq, you do not stop that convoy. If a little kid runs in front of your truck, you are under orders to run him over instead of stopping your convoy. This is the policy that's set in how to deal with people in Iraq.

As Gen. Stanley McChrystal, then the senior American and NATO commander in Afghanistan, said in March 2010: "We have shot an amazing number of people, but to my knowledge, none has ever proven to be a threat."

Aerial Bombardment

One of the most significant legacies of World War II has been the bombing of civilians. This new approach to war brought the front lines much closer to home while allowing those doing the killing to be located too far away to see their victims.

For the residents of German cities, survival "beneath the bombs" was a defining characteristic of the war. The war in the skies erased the distinction between home and front, adding "air terror psychosis" and "bunker panic" to the German vocabulary.

A U.S. pilot in the War on Korea had a different perspective:

> The first couple of times I went in on a napalm strike, I had kind of an empty feeling. I thought afterward, Well, maybe I shouldn't have done it. Maybe those people I set afire were innocent civilians. But you get conditioned, especially after you've hit what looks like a civilian and the A-frame on his back lights up like a Roman candle—a sure enough sign that he's been carrying ammunition. Normally speaking, I have no qualms about my job. Besides, we don't generally use napalm on people we can see. We use it on hill positions or buildings. And one thing about napalm is that when you've hit a village and have seen it go up in flames, you know that you've accomplished something.

Proponents of aerial bombing have argued from the start that it could bring a faster peace. This has always proved false, including in Germany, England, and Japan. The idea that the nuclear destruction of two Japanese cities would change the Japanese government's position was implausible from the start, given that the United States had already destroyed several dozen Japanese cities with firebombs and napalm.

Asia-Pacific Journal editor Mark Selden explains the importance of this horror to the decades of U.S. war-making that would follow:

> Every president from Roosevelt to George W. Bush has endorsed in practice an approach to warfare that targets entire populations for annihilation, one that eliminates all distinction between combatant and noncombatant with deadly consequences. The awesome power of the atomic bomb has obscured the fact that this

strategy came of age in the firebombing of Tokyo and became the centerpiece of U.S. war-making from that time forward.

The damage of our wars outlasts the memories of elderly survivors. We leave landscapes pock-marked with bomb craters, oil fields ablaze, seas poisoned, groundwater ruined. We leave behind—on the land and in the bodies of our own veterans—Agent Orange, depleted uranium, and all the other substances designed to kill people quickly but carrying the side effect of killing people slowly. Since the United States' secret bombing of Laos ended in 1975, some 20,000 people have been killed by leftover unexploded ordnance. Even the War on Drugs begins to look like the War on Terror when the spraying of fields with herbicides renders entire regions of Colombia uninhabitable.

War on Land: Toxic Burdens and Military Exercises

Susan D. Lanier-Graham (2017)

One of the first effects of military buildup is a relaxation of environmental standards. Prior to World War I, the United States had begun enacting legislation aimed at protecting the environment and conserving natural resources. Under the administration of Theodore Roosevelt (1901–9), 140 million acres were added to the National Park System. Legislation protecting land and water was passed. Big game and bird preserves were established. As we prepared for World War I, much of the legislation was overturned or ignored when claims of patriotism overrode environmental controls. National parks were opened to grazing, logging, hunting, and agricultural development that had been restricted in the early 1900s.

The role of preparing for war changed dramatically in the post–World War II years. The onset of the Cold War brought daily military rehearsals for the big war that everyone hoped would never come but decided to prepare for anyway. The military became a fixed part of American life—an entire industry was built up around military preparations for war. By the end of the Cold War, the Pentagon controlled 1,246 military bases (both in the United States and on foreign soil), 17 nuclear warhead production sites (operated by the Department of Energy), and 12 chemical weapons production and storage facilities.

While the military has begun taking precautions to contain and reduce toxic substances, the damage has already accumulated over more than fifty-five years. Denver's Rocky Mountain Arsenal has been dubbed "the most toxic square mile on Earth." All of the toxic byproducts from the production of mustard gas, napalm, incendiary weapons, and other types of munitions were dumped on the land for decades. Toxin ponds are lethal to wildlife, and yet poisons leaked into the groundwater for years before officials admitted there was even a problem at the arsenal.

Estimates are that more than 20,000 present and former government sites are contaminated with toxic substances. According to the nonprofit Center for Defense Information, the cleanup bill could reach or exceed $150 billion. No matter what the actual cost of the cleanup, the detriment to the environment will most likely never be fully assessed or reversed.

Throughout the Cold War, industrial wastes were routinely dumped into the local creeks at the Cherry Point Marine Corps Air Station in North Carolina. Today, these creeks contain high levels of mercury and lead. Local fish show signs of contamination and reproductive problems.

At Maryland's Aberdeen Proving Ground (formerly known as the Edgewood Arsenal), military and civilian personnel dumped arsenic, cyanide, napalm, white phosphorus, and other chemicals at a site that sits in close proximity to state-designated Critical Habitat Areas, a National Wildlife Refuge, and Chesapeake Bay.

Tinker Air Force Base, just outside Oklahoma City, was built in a major drainage basin atop central Oklahoma's only underground aquifer. Tinker AFB, a repair depot for aircraft weapons and engines, was contaminated with solvents such as trichloroethylene and heavy metals like hexavalent chromium. According to a 1991 report to Congress, the contamination covered 220 acres and included 6 landfills containing 1,705,000 cubic yards of industrial and sanitary waste.

In San Francisco, groundwater at the Hunters Point Naval Station Annex—a Navy shipyard originally built in 1869—tested positive for benzene, PCBs, toluene, and phenols. Studies also discovered higher-than-normal levels of heavy metals in offshore sediments.

Perhaps the worst case of unexploded ordnance was found at the Army's Jefferson Proving Ground in Indiana. A munitions test area since 1941, the military test-fired an estimated 23 million rounds of ammunition at the site. As many as 1.5 million of those test rounds remain unexploded, littering the facility's 55,000 acres with munitions—some reportedly buried 25 feet underground. Because unexploded munitions are not considered toxic substances, the cleanup of these munitions is not covered under the federal Resource Conservation and Recovery Act that pays for recovery costs at other contaminated sites.

American military bases worldwide are further sites of contamination. As far back as 1990, the army identified 358 contaminated sites at its European bases—the result of fuel leaks, solvent spills, hazardous substances dumped in landfills, and the ammunition expended at various firing ranges.

The United States is not the world's only military polluter. Levels of pollution generated in the former Soviet Union's military-industrial complex far surpassed that created by the United States. Without even the limited controls found in United States, the Soviets freely poured contaminants into the biosphere for five decades.

Military Exercises

Preparations for warfare involve increased levels of war exercises. One of the major differences before and after World War II was in the amount of space required for an army (or navy or air force) to train. In World War II, only 4,000 acres were necessary for full-scale tank and infantry maneuvers. Today, army officials claim they need 80,000 acres to carry out similar exercises.

During World War II, troops under Gen. George S. Patton held tank maneuvers in the Southern California desert. Today, those tracks still remain visible. In most areas, only about 35 percent of the vegetation has recovered.

Armored tanks continue to be a major cause of environmental damage. Tanks destroy vegetation, wildlife habitat, and wildlife. U.S. exercises held in Germany—including the massive "Reforger" exercises—caused hundreds of thousands of dollars in damage to the German countryside.

Reforger assembled large numbers of U.S. troops—many shipped in from National Guard bases in the States—to conduct field training in the German countryside. At its peak in 1988, Reforger involved 97,000 NATO soldiers (including 75,000 Americans) and 7,000 tracked vehicles. Thousands of troops engaged in maneuvers for approximately two weeks, tearing up fields and roads. In the damp German countryside, the combination of wheeled vehicles and tanks turned fields into giant mud traps. These maneuvers were not restricted to military bases—they extended into meadows and fields, over planted crops, and into nearby forests.

Local opposition to these damaging maneuvers became so intense that the Reforger exercise scheduled for 1989 was cancelled. The last Reforger exercise was conducted in May 1993. Unfortunately, in 2015, the growing tension between the United States/NATO and Russia prompted the Pentagon to dispatch U.S. Army tanks and Air Force planes to participate in a new era of military exercises in Europe under the banner of Operation Atlantic Resolve.

War on the Sea: Islands under Siege

Koohan Paik, International Forum on Globalization (2017)

In this age of ecological breakdown, pockets of wondrous biodiversity still survive in the vast Pacific Ocean. The Gulf of Alaska teems with a multitude of whale species; Southeast Asia's "Coral Triangle" boasts 500 species of coral; the Great Barrier Reef of Australia, the Galapagos Islands, and the seven-mile-deep Mariana Trench are still fairly intact. But these marvels may soon be wiped out by unchallenged trends in global militarism.

Widespread military exercises, defense-industry profiteering, and base-building (mostly by the United States) are wreaking irreversible destruction on coral reefs and other ecosystems, even without active war.

It is true that, for decades, deleterious war games have taken place on military range complexes spanning from Asia's east coast to the west coast of the Americas, and points in between. However, the scale and capacity for destruction has never been so immense as it is now. It's as if military activities have been suddenly "supersized." The U.S. Navy estimates that between 2016 and 2020, naval exercises in the Gulf of Alaska will kill over 180,000 marine mammals. In Hawaiian waters, it is estimated that 9.6 million marine mammals will be injured or killed over the same period.

But most galling is the new, fraudulent manner in which the United States has come to gain control of a whopping nine million square nautical miles of the Pacific Ocean—an area double the size of all fifty states. (This chicanery goes entirely unmentioned by reporters or politicians, so the public remains oblivious.)

The United States started claiming huge swaths of the Pacific about a decade ago, in anticipation of the threat of a rising China competing for finite resources and regional hegemony. The sweeping dominion of the United States took the form of "range complexes," slated for military practice, and "marine monuments," supposedly intended for environmental protection.

The first marine monument was designated in 2006, just before George W. Bush left office. He designated the Northwest Hawaiian Islands as the Papahanaumokuakea Marine National Monument. Environmentalists cheered this supposedly conservationist move. Ten years later, they cheered again, when Obama doubled the size to over a half-million square miles. What they didn't realize was that, in one fell swoop—without public participation or scrutiny—Bush and Obama were paving the way for militarizing vast tracts of the Pacific.

When commercial enterprise is banned, it turns out that a marine monument can easily morph into a military "range complex." This was the case with Papahanaumokuakea Marine Monument, which overlaps with the Northwest Hawaiian Islands Range Complex. Commercial and indigenous fishing are off-limits, but torpedoes, sonar, and all manner of detonations can blast with impunity. For example, the Environmental Protection Agency's "allowable" limit on cyanide is one part per billion parts of water. Nonetheless, the cyanide discharge from a single torpedo is in the range of 140 to 150 parts per billion.

The Pentagon insists that these war simulations are required to ensure military preparedness. But for the whales, turtles, dolphins, coral, sea sponges, snails, anemones, reef fish, sea urchins, and thousands of other diverse and rare species, living in a range complex is no "simulation." For them, it's real war, all the time. "Marine monument" status doesn't protect them.

A similar scenario took place when the Pacific Remote Islands Marine National Monument was expanded to include the Marshall Islands, infamous for its atomic-testing legacy. The new status has not stopped missiles and hypersonic aircraft from scattering shrapnel into the Marshalls' Kwajalein lagoon. Apparently, the real function of the "marine monument" designation is to introduce, without controversy, U.S. jurisdiction over the open seas.

Yet another example is supremely wild Pagan isle, within the Mariana Trench Marine National Monument in the Western Pacific. Pagan is a kind of "Noah's Ark," a miraculous habitat where populations of rare birds, snails, insects, plants, and animals thrive. Yet, the Pentagon now is proposing "full spectrum" military exercises on Pagan. That would mean year-round amphibious attacks, bombing, torpedoes, underwater mines, and other detonations from the air, from the sea, and from the ground, bombing the eighteen-square-mile island out of recognition. Nearby Tinian Island is also slated for live-fire training.

So much for "marine monument" protection. The designation is a fraud. In fact, three of the four U.S. "marine monuments" overlap with "range complexes"; the one that doesn't is the Rose Atoll Marine Monument. Even so, that monument, located in the Southern Hemisphere east of American Samoa, still serves a geostrategic function: to physically block China's access to its newly established South American interests.

Base-Building and Resistance

Base-building is another ecocidal activity on the rise. There are already more than 400 official U.S. bases throughout the Asia-Pacific region. Meanwhile, client states, such as Japan and South Korea, have been enlisted to build their own installations that would effectively encircle China with missiles. New bases have been built on Jeju Island, in Korea, and in Japan's Ryuku chain—on Okinawa, Miyakojima, Amami Oshima, Ishigaki, and Yonaguni, only seventy miles from Taiwan. These bases turn peaceful islands into strategic targets.

Islanders determined to protect their homes have not remained silent. On Okinawa, ferocious opposition has significantly delayed the two-decade-old plan to build a U.S. base at lovely Oura Bay. Sadly, the Japanese government successfully installed dozens of twenty-ton concrete blocks atop coral reefs there. However, Okinawa's anti-base governor Takeshi Onaga has joined the activists on the ground, filed lawsuits, and made three visits to Washington to personally explain Okinawa's intractable stance.

And on Jeju Island, in South Korea, a Navy base (designed to port Lockheed Martin Aegis-missile destroyers) has been built at Gangjeong village, adjacent to a UNESCO Biosphere Reserve. The base construction has destroyed a unique rocky wetlands and also a rare coral ecosystem that was home to Korea's last remaining dolphin pod.

Against the stars-and-stripes backdrop of expanding range complexes, marine monuments, and base-building, other nations are also contributing to the demise of a healthy Pacific. This resource-rich sea—framed by China, the Philippines, Malaysia, Brunei, and Vietnam—has become the region's most disputed territory. At one point, the Philippines became so distressed by China's incursion into the area's Spratly Islands that it agreed to allow U.S. troops and ships to return to its former bases, from which they were passionately evicted in the 1990s.

In June 2015, over 11,000 American and Filipino troops (double the number of soldiers from previous years) participated in joint naval war games on beautiful Palawan Island. But that was under President Benigno Aquino. Newly elected president Rodrigo Duterte has called off any new U.S. military exercises in his nation and has engaged in diplomatic talks with China to defuse the volatile scenario.

For its part, China has built seven artificial islands, three of which are militarized, smack dab in the middle of the Spratlys. The islands, with a total area of 2,000 acres, are built from dredged and crushed coral spread over what were once some of the world's most vibrant reefs—now certainly dead.

Ecocide, for Games and Profit

Such fever-pitch tensions are actually viewed by the Pentagon as a window of opportunity. In 2015 and 2016, Defense Secretary Ashton Carter completed several barnstorming tours of Asia to solidify alliances with nations seeking U.S. muscle. This resulted in an unprecedented onslaught of joint naval exercises that included the following partnerships: U.S.-Australia-Japan-New Zealand; U.S.-Philippines; U.S.-South Korea; U.S.-Japan; and U.S.-India-Japan. While in India, Carter signed a ten-year agreement that, among other things, recruits India's top corporations to cheaply build jet engines, aircraft carriers, and other high-end armaments.

Carter's visit also inspired new war-games partnerships between U.S.-ally nations, including a partnership between India and Singapore. (The United States and India conduct at least fifty joint naval exercises per year.) In 2016, for the first time in history, China and Russia conducted joint military training in the Sea of Japan and the South China Sea. If this ecocidal saber-rattling continues, it would effectively wipe out 10 percent of the world's global fish supply, which, according to the UN Global International Waters Assessment, is the volume of catch produced in this region.

In 2015 and 2016, the Pacific region experienced scores of naval and land exercises involving thousands of troops and hundreds of armed vessels. In addition to forces from the United States, these joint exercises drew soldiers and sailors from Australia, Indonesia, Japan, Malaysia, New Zealand, and the Philippines. The Pacific region also has served as a staging area for troops from Bangladesh, Brunei, Cambodia, France, India, Pakistan, Sri Lanka, Thailand, Timor-Leste, and Vietnam. Here are a few of the larger exercises:

2015

July: "Talisman Sabre" took place over 10 days on Australia's beautiful north coast with 640 officers from New Zealand and Japan—backed by 33,000 soldiers, 21 ships, Osprey and Cobra helicopters, jet fighters, three submarines, amphibious assault ships, and more than 200 aircraft. One goal of Talisman Sabre: to stage a naval blockade of sea lanes that could prevent oil from reaching China.

August: Russia and China staged massive drill "in response" to U.S. exercises. Joint Sea 2015 II involved sixteen warships, two subs, aircraft, and helicopters in the Sea of Japan.

September: U.S. "Valiant Shield." A twelve-day multi-force exercise with nine warships, 180 aircraft, and 18,000 personnel off Guam and the Marianas.

2016

February: United States, Australia, Japan, New Zealand, South Korea, and the Philippines staged a multinational air power exercise on Guam. This largest "Cope North" exercise in history involved 3,000 personnel and 100 aircraft from the United States, Japan, and Australia.

March–May: United States and South Korea. Operation "Foal Eagle" involved 50 ships, submarines, aircraft, and 315,000 troops, double the number from 2015.

April: Some 5,000 U.S. soldiers joined more than 3,000 troops from Australia and the Philippines for Operation Balikatan, with air-land-and-sea strikes targeting the Philippine islands of Luzon, Panay, and Palawan.

July–November: United States, Bangladesh, Brunei, Cambodia, Indonesia, Malaysia, Thailand, Timor-Leste, and Vietnam staged land-and-sea combat exercises in the waters off Singapore.

August: United States and South Korea held annual Ulchi Freedom Guardian war games with live missile firings in the Sea of Japan. Approximately 50,000 South Korean and 30,000 U.S. troops engaged in command-and-control exercises that North Korea condemned as a "provocation."

September: China and Russia. A total of 18 ships and supply vessels, 21 aircraft, and more than 250 personnel convened for one week in the South China Sea for Joint Sea 2016 exercises to practice what China's Xinhua news service described as "air defense" and "island seizing" operations.

October: More than 1,900 U.S. and Philippine troops conducted amphibious landing exercises and live-fire training in Luzon between the Sulu Sea and the South China Sea. Other joint naval exercises involved: China and Thailand, India and

Mauritius, Philippines and Cambodia, and the United States, Japan, and India.

And then there is RIMPAC (Rim of the Pacific), the granddaddy of joint naval exercises. Part World Cup, part trade show, RIMPAC is the chance for 25,000 troops from 22 nations, 55 vessels, and more than 200 aircraft to gather every four years in Hawaii. For two weeks, they drop bombs, shoot missiles, set explosions, and sink aircraft carriers at the Papahanaumokuakea Marine Monument. And sell a missile or two.

Lockheed Martin has shown its capitalist foresight by moving into the undersea-mining-technology sector. The idea is to profit by selling missiles and destroyers to nations fighting for mineral-rich territories, and then sell the mining technology to whichever nation prevails. Lockheed Martin wins, both coming and going, while the creatures of the ocean perish either way.

U.S. corporations were not the only ones planning to profit from Obama's "pivot toward Asia." With Japan relaxing its postwar ban on defense manufacturing, Mitsubishi anticipates healthy sales of its new amphibious tank that operates three times faster than its U.S. counterpart. Its main selling point is that coral-crunching traction over reefs is not compromised by its unprecedented speed.

Another environmentally chilling development is the inauguration of a RIMPAC-like event focusing on amphibious warfare. Sponsored by the U.S. Marine Corps, the event's purpose is to develop procedures and technologies best suited to dominating the South China Sea. The list of participating countries is surprisingly diverse: Australia, Bangladesh, Cambodia, Chile, Canada, Colombia, France, Indonesia, Japan, South Korea, Malaysia, Maldives, Mexico, New Zealand, Peru, Philippines, Singapore, Sri Lanka, Taiwan, Thailand, Tonga, the United Kingdom, and Vietnam.

Our oceans, which are already suffering from overfishing and gyres of plastic waste, supply up to 80 percent of our atmospheric oxygen. Our reefs, the essential foundation for all marine life, are already dying due to warming waters and acidification. Yet war profiteers seem determined to bomb all life out of the Pacific. Such behavior is intolerable. Don't they know there are no winners on a dead planet?

Wars for Sand: The Mortar of Empires

Denis Delestrac (2016)

From Gallipoli to Normandy, some of history's bloodiest battles have been fought on the world's beaches. In the near future, however, some of our fiercest battles may be fought *because* of our beaches or, more correctly, because of the sand that shapes them. Like oil, gas, and all nonrenewables, sand (after water, the most consumed resource on Earth) is now on the front line of a growing global war over access to raw materials.

For most of us, sand conjures up images of relaxing days at the beach. But, sand's impact on our lives spreads far beyond the shoreline. Melted and transformed into glass, it sits on every shelf. It's also the source of silicon dioxide, a mineral found in cleaning products and detergents, in paper, dehydrated foods, hairspray, toothpaste, cosmetics, and an astounding variety of products we use on a daily basis.

The minerals extracted from sand form the foundations of our hyperconnected society, including microchips, without which our computers, credit cards, bank machines, cell phones, and many other devices would not exist. Sand is everywhere. It's in our household plastics and it's in our cars, trains, buses, and planes. It's in the lightweight alloys of the jet engines, the fuselage, the paint, and even the tires.

And for the last 150 years, sand—mixed with cement—has helped sculpt the contours of our urban landscape. Two-thirds of the world's buildings are made of reinforced concrete—a blend of cement, sand, and gravel. Because of its strength and low cost, concrete has become the world's dominant building material. The quantities used are astronomical. To build an average house takes 200 tons of sand. For a school or a hospital, around 3,000 tons. Each mile of highway devours 48,000 tons. And to build a nuclear plant? The estimate is about 12 million tons.

Sand constitutes the largest volume of solid material extracted on the planet, with annual consumption totaling 15 billion tons. With beaches disappearing, most sand we now use is dredged from beneath the ocean.

While modern civilization has depleted the available sand, it has also interrupted the process of sand formation. Sand is formed from the actions of glaciers and water breaking apart rocks in the high mountains and is ultimately carried to the ocean with river sediments. But there now are at least 845,000 dams on the planet and it's not only water they are holding back. One-quarter of the planet's sand reserves lie trapped behind these dams. Any sand that makes it beyond the dams often encounters river dredging.

Dubai is an astonishing example of our voracious appetite for sand. Within a few decades, this fishing village has morphed into a Mecca of modern architecture—a sandbox for developers where no fantasy is too grandiose. But Dubai's delusion of grandeur swallows up a lot of sand.

Flying high on a seemingly endless supply of oil money, Dubai embarked on an extravagant construction project called "World"—an artificial archipelago of 300 islands, designed as a map of the world. This self-proclaimed "eighth wonder" cost more than $14 billion and devoured more than 150 million tons of sand—all dredged from Dubai's coastline.

Today Dubai's World is a mirage. The worksite was abandoned at the onset of the 2008 financial crisis, and the deserted islands now parch in the sun. Overdevelopment has liquidated Dubai's natural sand resources and, *sans* sand, the construction-addicted emirate has reached an impasse. (In order to finish building the Burj Khalifa, the world's tallest building, Dubai had to import sand from Australia. The emirate continues to import sand, even while 90 percent of the Burj Khalifa's apartments remain vacant.)

But how can this be when Dubai is surrounded by sand? Michael Welland, a geologist and author of the book *Sand*, explains: "Desert sand is the wrong kind of sand for building artificial islands. Why? Because desert sand . . . [is] typically very round and very smooth. . . . To build an island, you need sand that is more angular [because] . . . it's rougher-edged sand that naturally sticks together."

Thanks to its finite nature and overexploitation, the market for sand is booming with profits tripling in twenty years. But this doesn't bode well for the health of the oceans. For when ships scour the sea bottom for sand, everything on the seafloor is dredged as well—including crustaceans, fish, plants, and coral.

Singapore's land mass has increased 40 percent in the past 20 years with 130 square kilometers [50 square miles] converted from coastal water to land using dredged sand. The Singaporean government is the largest customer of illegal sand traders.

One of the most visible impacts of the sand trade is the disappearance of twenty-five islands off the coast of Indonesia. Meanwhile, dozens of barges filled to the brim can be seen unloading sand on a daily basis. Where does this sand come from? Thanks to local trafficking networks, dealers with false identities, working for fictional companies, continue to remove sand from beaches in Cambodia, Indonesia, Malaysia, and Vietnam.

In India, the sand mafia—the country's most powerful criminal organization—uses blackmail and violence to enforce loyalty. Its influence reaches far beyond the extraction sites. Under the eyes of corrupt authorities, sand pirates ply their trade in broad daylight at more than 8,000 dredging sites along the coasts and riverbanks of the subcontinent.

The illegal theft and sale of sand affects every continent. Tons upon tons of sand are removed every day from the beaches of Africa, Asia, and the Caribbean. This pillaging goes on nonstop, year-round.

Much of this sand is sold directly on the construction sites of unscrupulous developers. But there is a risk. If not properly rinsed in fresh water before being added to cement, sand mixed with salt water becomes highly corrosive, which could eventually cause buildings to collapse.

In the Maldives, where reserves of sand derived from coral are dwindling fast, the islands are eroding at an alarming rate and the future of a whole population is under threat. The Maldives' most vulnerable islands now stand deserted after their residents fled to larger, better-protected islands. The overpopulated capital is building new houses to meet the demand but, ironically, the resulting construction boom is fed by sand removed from nearby lagoons—the very sand that is supposed to protect the islands from steadily rising seas.

We've never built on such a scale but, at the same time, housing has never been less available. One-third of the urban population lives in slums, while sprawling "ghost cities" and empty apartments are built all over the world. Entire airports have been built, without seeing a single passenger. In China, 65 million flats are empty, yet the construction industry is flourishing.

Like the automobile industry's addiction to oil, the construction industry is addicted to sand. But the leading consumer of sand is the state—a massive builder of infrastructure like highways, bridges, ports, etc. (The next time you look at a network of highways, pause to remind yourself that "roads are made from the world's beaches.")

Around the world, the planet's beaches are shrinking at an accelerating rate, largely due to sand dredging. By 2025, three-quarters of the world's inhabitants will live near the ocean, and those thin ribbons of sand that surround the continents are feeling the pressure. Unregulated dredging has transformed once-healthy beaches into lunar landscapes. Between 75 and 90 percent of our beaches are shrinking—and the trend is accelerating. If nothing is done, by 2100, the world's beaches could be history.

Grain after grain, beaches slowly erode, mute victims of decades of human interference. Overexploitation of sand—combined with dams and bad management of shorelines—is creating an explosive mix for the planet's beaches.

War at a Distance: Long-Range Missiles

Jon Letman (2014)

Whale-watching tourists sailing along Kaua'i's famed Na Pali Coast have no idea they are sharing the scenic waters with defense contractors Boeing, Raytheon, Lockheed Martin, Northrop Grumman, and General Atomics, among others.

But just as Kaua'i is a major center for biotech companies, so is it a critical, albeit under-reported, testing and training hub for the U.S. military, the Defense Advanced Research Projects Agency (DARPA), NASA, the Missile Defense Agency (MDA), and Sandia National Laboratories, which all conduct programs at the Pacific Missile Range Facility (PMRF).

Unlike O'ahu, Kaua'i's military presence goes largely unseen. Yet PMRF, which has nearly seventy-five-year-old roots as an Army landing site, occupies a 7.5 by .75-mile-wide swath of coast at Barking Sands on the edge of the Mana Plain. This once marshy wetland, long ago drained to grow sugar cane, is today home to GMO crop fields that form a buffer around PMRF.

For most tourists and even many Kaua'i residents, PMRF remains *terra incognita*. Although PMRF is covered in the local *Garden Island* newspaper and the *Honolulu Star-Advertiser*, it's rarely mentioned in national or international media—unless there's a major launch, like the May 22, 2014, Aegis Ashore ballistic missile defense (BMD) test.

In May 2014, I had the chance to spend a few hours touring PMRF with its commanding officer, Captain Bruce Hay, as my guide. After a cafeteria-style lunch at PMRF's Shenanigans cafe (during which Capt. Hay promised our tour would be "open kimono"), we were led to his car—a Chevy Malibu as blinding white as his uniform. We piled in and were taken on a short drive to a 135-foot-long rocket launch rail resembling a gray steel bridge jutting out over the sand.

The captain described how the launch was designed to fire the Space-Borne Payload Assist Rocket, also called Super Strypi—a collaborative project of Sandia National Laboratories, Aerojet Rocketdyne Corp., the University of Hawai'i, and PMRF. The main function of its first mission will be to demonstrate how to inexpensively deliver a 300-kilogram (662-pound) payload into low-earth orbit. For the project's partners, Super Strypi represents not only a technical achievement but also an educational opportunity as private and public sectors push to advance Hawai'i's position in the world of the aerospace and defense industries.

Launching a rocket—any kind of a rocket—at PMRF is, for Hay, "pretty neat." Over the course of the two-and-a-half-hour tour, he repeatedly expressed his enthusiasm for involving children in launches (as observers), whenever possible. It's all part of PMRF's—and more broadly, the aerospace and military defense industry's—enthusiasm for supporting STEM (Science, Technology, Engineering, and Math) education.

Not Just Kaua'i

Contrary to popular belief, PMRF is not limited to Kaua'i's Mana Plain. The range also includes "small installation" sites at Koke'e State Park, Makaha Ridge, Kamokala Ridge, Port Allen, Mauna Kapu on O'ahu, Pōhakuloa Training Area on Hawai'i Island, and the privately owned island Ni'ihau, which is home to a "perch site" comprised of a helicopter pad, electronic warfare equipment, and surveillance radar.

Less than two dozen miles south of Ni'ihau stands Ka'ula, a steep offshore islet inhabited by bird colonies. Like Kaho'olawe island and Pagan island in the Northern Marianas, part of Ka'ula is being used for "inert air-to-surface weapons" testing. Unlike Kaho'olawe and Pagan, however, there is effectively no protest or even knowledge that this island is being used.

As we drove from one site to the next, Hay pointed out that Ka'ula's ownership is in question. "Is it the Federal Government? Is it the Department of Defense? Is it the State of Hawai'i? I don't know. I'm not a lawyer—thank God," he quipped.

One of PMRF's greatest assets, Hay noted, is the fact that it has a whole lot of nothing. That "nothing" is, in fact, over 2.1 million square miles of "extended range"—roughly the size of the continental United States west of the Mississippi including Baja, Mexico. This is in addition to 1,100 square miles of "instrumented sea range" and 42,000 square miles of "controlled airspace."

From Hay's perspective, conducting testing and training west of Kaua'i means the military is not impacting shipping lanes or civilian air traffic. Critics, however, are increasingly challenging environmental impacts, particularly the use of high-frequency sonar and its alleged impact on marine life from deepwater corals to large marine mammals.

Out in that vast watery "nothingness," PMRF operates MATSS (Mobile-At-Sea-Sensor System)—a barge loaded with antennae, telemetry tracking dishes, and equipment used to support BMD testing.

PMRF touts itself as an important economic driver on Kaua'i, proudly describing its role in employing local people. The base employs around 770 contractors, tenants, and other service providers, along with 140 civilian and 87 military personnel (officers/enlisted). Currently, about 150 people live on base in some 50 modest housing units.

Although PMRF has many of the amenities you'd find in a small town (gas station, car wash, barbershop, post office, an outdoor theater, Subway sandwich outlet, parks, and recreational sports facilities), it doesn't feel like your average civilian town. Maybe it was the Regulus cruise missile mounted on display, but the place just feels like a military base, which, of course, it is.

Driving along PMRF's almost carless roads, we slowed down to look at roadside tanks. Hay said the dummy tanks—called "composites"—are used for pilot training. The tanks aren't fired at with live ammunition, but they can provide a realistic training object, particularly when equipped with heat generators that simulate a "live" target.

Next we stopped at an unremarkable beige building, which Hay identified as his primary working headquarters—the Range Operations Complex or "the Roc." Hay led us to a Standard Missile Three (SM-3) displayed on a mount behind a plaque that reads *"Ad Astra Per Aspera"* ("To the stars through adversity").

The SM-3 (manufactured by Raytheon, which describes it as "the world's only ballistic missile killer deployable on land or at sea") is designed to "engage non-air-breathing ballistic missile targets." In other words, the SM-3 is intended to be fired at an incoming enemy missile and destroyed by sheer kinetic force. It's the "kill vehicle," Hay explained—gesturing to the rocket's twenty-one-inch tip—that matters most. "It's pretty neat to think we are hitting a bullet with a bullet," Hay said, referring to the 21.5-foot-long white priapic missile behind him.

Hay described ballistic missile defense as "very successful, in the low nineties," but a *Los Angeles Times* investigation found that the Boeing-manufactured $40 billion Ground-Based Midcourse Defense System, tested at PMRF-partnering facility Vandenberg Air Force Base in California, was "unreliable, even in scripted tests." One physicist at Lawrence Livermore National Laboratory called the system's test record "abysmal."

PMRF has a long history of supporting missile defense testing. One of the most high-profile systems being tested today is Aegis Ashore. Essentially identical to the BMD system deployed on Aegis naval destroyers, Aegis Ashore is designed for use on land.

A product of Lockheed Martin, Raytheon, the U.S. Navy, and the Missile Defense Agency, Aegis Ashore was championed by the late senator Daniel Inouye and was slated for deployment in Romania in 2015 and Poland in 2018.

As Capt. Hay pulled up in front of a dull, white building topped with radar antenna equipment behind a high black fence topped with barbed wire, he said, "If you're familiar with ships, [this] looks a lot like a

Ticonderoga Class (guided missile) cruiser." He explained that operating the Aegis on a ship requires 300 to 400 people. "Take a stab at how many sailors run this facility," he said.

"Fifteen?" I guessed.

"Less, but not much. Twelve sailors and a few contractors," he replied. According to Hay, the building alone (Aegis Ashore Missile Defense Test Complex) cost $60 million. Once fitted with radar and other equipment, the cost soared to $700 million. On a ship, Hay said, that would be around $2 billion—even before crew salaries and fuel.

The first Aegis Ashore test took place at PMRF just three days before my visit and made news for the mysterious vapor trails it left behind. But the real news no one seems to talk about is how Kaua'i stands at the center of a BMD system that Washington insists is to protect Europe from Iran but whose deployment has been repeatedly and angrily criticized by Russia.

When I asked Capt. Hay who Aegis Ashore is intended to defend against, he declined to name names, instead saying: "Think of all the antagonists all throughout Europe and the Middle East." And, Hay added, "the former Soviet Union wasn't too crazy about it, for obvious reasons." I did not ask how the United States would respond to a Russian BMD deployment in Mexico, Canada, or Cuba.

PMRF is "not just ballistic missile testing," Hay pointed out in an interview with the *Garden Island* newspaper. "We're doing big things for very important people all across the globe."

"Big things" presumably include supporting Unmanned Aerial Vehicle (UAV) or drone testing and training for systems like the MQ-Reaper, MQ-1 Predator, and the high-altitude capable ALTUS II, as well as NASA research aircraft and other UAV systems like the Coyote and the Cutlass V. PMRF has also seen visits by fighter jets and the V-22 Osprey, a hybrid aircraft with a checkered safety record and the object of ongoing protests in Okinawa, where it is deployed.

Doing "big things for very important people" also means hosting Sandia National Laboratories' Kaua'i Test Facility (KTF). Sandia, a wholly owned subsidiary of Lockheed Martin, is one of the United States' three primary nuclear weapons labs. KTF was established on Kaua'i as a tenant inside the PMRF in 1962 to support Operation Dominic, which included a series of thirty-six high-altitude nuclear weapons tests over the Pacific.

In November 2011, KTF was the launch site of the Advanced Hypersonic Weapon (AHW), a missile that is intended to fulfill the goal of a "Prompt Global Strike," a directive that would enable the United States to bomb anywhere on Earth in under sixty minutes. In the November 2011 test, the AHW was fired from Kaua'i, arriving at the Reagan Test Site on Kwajalein Atoll in the Marshall Islands, about 2,500 miles away, in 30 minutes.

Since its inception in 1962, KTF has supported 443 rocket launches from multiple sites (as of January 2017), making it—and its host PMRF—major players in a militarized Pacific.

When asked directly if nuclear weapons or components of nuclear weapons have ever been stored or passed through PMRF, a spokesman replied, "Per Department of Defense policy, all U.S. military installations can neither confirm nor deny the presence of nuclear weapons."

In the summer of 2014, twenty-three nations converged on Hawai'i for the RIMPAC (Rim of the Pacific) biennial maritime exercise. While the war games included countries as diverse as India, China, Singapore, Norway, Tonga, and Japan, the participants page of RIMPAC's website shows the U.S. military was overwhelmingly represented.

RIMPAC provides a chance to show off the latest military technology, gadgetry, and test systems like drones (during RIMPAC 2012, an AeroVironment sub-launched "kamikaze" drone was tested) and practice live-fire sinkings of decommissioned ships in an exercise called SINKEX (Sinking Exercises).

Besides this, RIMPAC provides a realistic setting for urban combat training, amphibious landings, underwater sonar training, and a host of other military exercises.

Kaua'i may be only 35 square miles larger than the city of Phoenix (with less than 5 percent of its population) but, thanks to PMRF, it plays an outsized role in America's ability to wage wars; control the seas, skies, and space; and ensure that the U.S. military juggernaut can continue in its quest to maintain Full-Spectrum Dominance. Like the Ronald Reagan Ballistic Missile Defense Test Site in the Marshall Islands and Vandenberg Air Force Base in California, PMRF is a key spoke in the military's missile-testing arsenal.

As I drove away, a hard rain began to fall and I reflected: If PMRF is our base, then it is also our *kuleana* (responsibility) to understand what goes on inside and to make the connections between it and events around the world. Militarism and war do not take place in a vacuum. What happens here affects people around the world. It is incumbent on us to closely follow what our base is doing beyond the occasional headline rocket launch.

We need to understand that our base impacts lives in faraway places—from the dun-colored hills of Afghanistan and the war-torn cities of Syria and Iraq to the shallow blue lagoons of Micronesia's coral atolls and the gritty urban landscapes across the United States where many veterans end up after war.

War in Space: Astro-Imperialism

Karl Grossman (2017)

Vision for 2020, a report of the United States Space Command (USSC), declared that the United States sought to deploy weapons in orbit to "control space." This 1996 report spoke of dominating "the space dimension of military operations to protect U.S. interests and investment."

U.S. military leaders have been blunt through the years in describing plans for space warfare.

"It's politically sensitive, but it's going to happen," said Gen. Joseph Ashy, USSC commander-in-chief. In a 1996 article in *Aviation Week and Space Technology* titled "Future Combat Missions in Space," he stated: "Some people don't want to hear this, and it sure isn't in vogue, but absolutely—we're going to fight in space. We're going to fight from space and we're going to fight into space. That's why the United States has development programs in directed energy and hit-to-kill mechanisms."

As then assistant secretary of the Air Force for Space Keith Hall declared: "With regard to space dominance, we have it, we like it, and we're going to keep it."

Phillips Laboratory, a major Air Force contractor, has proudly described its mission as "helping control space for the United States."

There's been keen interest by U.S. administrations—the Reagan administration with its "Star Wars" plan a leading example—in placing weapons in space. That has alternated with some administrations more or less opposed—the Obama administration, for instance. On the other hand, the George W. Bush administration (with Richard Cheney as secretary of defense) was gung ho for the weaponization of space.

Under the Trump administration, it is highly likely that there will be a move by the United States to deploy weapons in space. If this happens, it will be profoundly destabilizing, setting off an arms race in space and, likely, leading to war in space.

The deployment of weapons in space is intimately linked to the use of nuclear power in space. The Reagan "Star Wars" program was predicated on orbiting battle platforms with onboard nuclear reactors providing the energy for their weaponry.

As Gen. James Abramson, director of the Strategic Defense Initiative Organization, told a symposium on Space Nuclear Power and Propulsion in Albuquerque, New Mexico, in 1988, "Without reactors in orbit [there is] going to be a long, long light cord that goes down to the surface of the Earth." He went on to note that "failure to develop nuclear power in space could cripple efforts to deploy anti-missile sensors and weapons in orbit."

In his 1989 book, *Military Space Forces: The Next 50 Years*, military specialist John M. Collins described how orbiting nuclear reactors could power space-based lasers, neutral particle beams, and other weapons while ground-based reactors could provide energy to support military bases on the moon. Collins' book, which was commissioned by the U.S. Congress, hailed the "unilateral control of space, which overarches Planet Earth, all occupants and its entire contents" and observed that "possessors of that vantage position could overpower every opponent."

The U.S. position in regard to weaponizing space is a violation of the intent of the Outer Space Treaty of 1967, the landmark agreement that sets space aside for peaceful purposes. Signed by ninety-one nations—including the United States, China, the United Kingdom, and the former Soviet Union—it decrees: "The exploration and use of outer space, including the moon and other celestial bodies, shall be carried out for the benefit and in the interest of all countries" and states that no nation shall "place in orbit around the Earth any objects carrying nuclear weapons or any other kinds of weapons of mass destruction."

Moreover, there have been repeated attempts since 1985 to pass a UN resolution titled "Prevention of an Arms Race in Outer Space" (PAROS) to assure that space "shall be used for peaceful purposes."

The United States has continually balked on this measure. For example, on November 1, 1999, 138 member nations of the United Nations voted for this resolution. The United States abstained. Canada, Russia, and China have been leaders in urging passage of the PAROS treaty.

With the arrival of the Trump administration, a renewed drive to weaponize space appeared in the offing. In a post-election report in November 2016 in *Roll Call*, Representative Trent Franks (R-AZ) predicted "a big payday is coming for programs aimed at developing weapons that can be deployed in space." The article was headlined "Under Trump, GOP to Give Space Weapons Close Look."

A November 22, 2016, article in *Blasting News* forecast that the new administration would be looking at "space-based weapons that could strike targets on Earth." This could include so-called "Rods of God"—tungsten projectiles fired from space.

The administration's $1.15 trillion "America First" budget—released on March 16, 2017—noted that Trump's proposed $52 billion increase in Pentagon spending "exceeds the entire defense budget of most countries." The key Defense Department programs listed in the budget document included: "American superiority . . . on land, at sea, in the air, and in space."

Prior to the release of the budget outline, Bruce Gagnon, coordinator of the Global Network Against Weapons and Nuclear Power in Space, founded in 1992, cited "some very disturbing initial recommendations that have been surfacing." The reported Trump administration plans "indicate the mindset that a full-blown war in space is in the thinking of some now coming to power," Gagnon warned. "The world does not need a new arms race in space—especially when we should be using our resources to deal with the real problems of climate change and growing poverty due to increasing economic divide."

In an interview at the University of Arizona on February 2, 2017, Prof. Noam Chomsky was asked about the Bulletin of Atomic Scientists' "Doomsday Clock," whose minute hand had just been moved forward to a mere thirty seconds to midnight ("midnight" representing a global nuclear holocaust). Did he share the organization's concerns about the risks of a Trump presidency?

"Trump has been very inconsistent on many things," Chomsky noted. "His statements are all over the map, but his personality is frightening. He's a complete megalomaniac. You never know how he's going to react." Chomsky reminded the audience that even a "first strike" is an act of suicide since it would trigger a "nuclear winter" that would destroy the world's crops and trigger global famine.

In his 1984 book, *Arming the Heavens: The Hidden Military Agenda for Space*, Prof. Jack Manno concluded that weapons, no matter how advanced, would never bring peace: "Only by eliminating the sources of international tension through cooperation and common development can any kind of national security be achieved in the next century."

THE MACHINERY OF MAYHEM

Every gun that is made, every warship launched, every rocket fired signifies, in the final sense, a theft from those who hunger and are not fed, those who are cold and are not clothed. This world in arms is not spending money alone. It is spending the sweat of its laborers, the genius of its scientists, the hopes of its children. This is not a way of life at all in any true sense. Under the clouds of war, it is humanity hanging on a cross of iron.

—Dwight D. Eisenhower

Fueling the Engines of Empire

Barry Sanders (2009)

While the U.S. military directed its "Operation Iraqi Freedom" (2003–10) solely against the Iraqis, no one—not a single citizen in any part of the globe—escaped its fallout. When we declare war on a nation, we now also declare war on the Earth—on the soil and plants and animals, the water and wind and people, in the most far-reaching and deeply infecting ways. War insinuates itself, like an aberrant gene, and, left unchecked, will eventually destroy Earth's entire ecosystem.

As we contemplate America in the opening years of the twenty-first century, we might reconsider George Washington's farewell warning that "overgrown military establishments . . . under any form of government, are inauspicious to liberty."

I write as a citizen, not a politician; as a layman, not a scientist; as an outsider from the academy, not an insider from the Pentagon. Most of the information that I present here the Department of Defense (DoD) withholds from the general public, makes intentionally obscure, folds inside arcane reports, and hides on hard-to-find governmental websites. The vital statistics for almost all armament—their type and number—remain highly classified. The number and kinds of vehicles and planes housed at 860 American bases in foreign countries also remain a mystery.

By my count, in 2005, the U.S. Armed Forces commanded the deserts and the neighborhoods of Iraq with about 30,000 vehicles. (According to its 2007 figures, the DoD inventory of non-tactical vehicles worldwide totaled 187,493, 13 percent based overseas.) These vehicles consist of hundreds of that most ubiquitous military mule, the Jeep, along with other familiar vehicles like up-armored Chevrolet Suburbans and Humvees.

Armor often means the difference between living and dying, and so the Army relies on heavily armored machines like the HMMWV (M1114), the Guardian Armored Security Vehicle M1117 (popular with MPs), the Cougar HEV Armored Truck, the LAV (light armored vehicle), the ICV (infantry carrier vehicle), the Stryker troop carrier, the High Mobility Multi-Purpose Wheeled Vehicle M1151, the Bradley Fighting Vehicle, the M-1 Abrams tank, and a behemoth called the Mine-Resistant Ambush-Protective vehicle, a thirty-ton armored bus for ferrying VIPs, and popularly known as the Rhino Runner. Military brass covet the Rhino Runner, for it holds the distinction as the most heavily armored, safest vehicle ever manufactured. These are but a few of the vehicles in the Army's inventory.

A United Nations environmental report about the first Gulf War points to the damage inflicted by seventy-ton tanks like the M-1 Abrams on the ecology of the desert: "Approximately 50 percent of Kuwait's land area has had its fragile soil surface destroyed as scores of tanks moved out of that country each day and headed for Iraq." Once the surface of the earth has broken apart, the wind has an easier job of eroding even more land mass.

The military—the Army, Navy, and Air Force—leads the world in its wide range of flying machines, like the fixed-wing A-10 Attack Jet, the F-16 Fighting Falcon, the F/A-18 and F/A-18 E Super Hornet, the CASA 212, a short-takeoff-and-landing transport aircraft, the B-52, B-1B, and B-2 bombers, the MQ-1 Predator, and rotary-wing helicopters like the Blackhawk, the CH 47 twin-rotor, the H-53E Sea Stallion, the CH-53D Super Stallion, the MH-53J Pave Low III, and the MH-53E Sea Dragon (the world's largest helicopter), along with dozens of other fixed- and rotary-wing aircraft, including headline-grabbing monsters like the F-22, B-52, F-111, and the F-117 Stealth bombers.

In all, I count forty-three different types of fighter planes, eleven attack planes, thirteen bombers, sixteen cargo planes, and nine different kinds of helicopters—all told, ninety-two different kinds of aircraft. Military brass also have their own private airline, the Air Mobility Command (AMC). For the most part, the Pentagon prefers to keep these planes off the record. The AMC consists of a fleet of long-range C-17 Globemasters, C-5 Galaxies, C-141 Starlifters, KC-135 Stratotankers, KC-10 Extenders, and C-9 Nightingales. As an additional perk for generals and admirals, the military provides seventy-one Learjets, thirteen Gulfstream IIIs, and seventeen Cessna Citation luxury jets for their private use.

Oil Is the Blood of War

Every day, these armored vehicles, aircraft, and luxury planes consume close to two million reported gallons of oil—a commodity that some critics of the war say we are fighting to protect. But such a contradiction should not seem so strange, for the business of the military—in times both of peace and war—is oil itself.

The United States must have adequate supplies of oil to maintain its position as the world's most prosperous nation. Of the Army's top-ten gas-guzzlers, only the M-1 Abrams tank and the Apache helicopter are combat vehicles. Ironically, the military needs most of its fuel-famished vehicles—along with a good number of its troops—for resupplying its vast fleet of fuel-dependent combat vehicles and fighter planes. These support vehicles consume over half the fuel in the battlefield. Fuel is the lifeblood of these vehicles. They require it in astonishing amounts, consume it with astonishing speed, and demand it with astonishing rapacity.

According to the U.S. Defense Energy Support Center (DESC), military fuel consumption for fiscal 2004 hit 144 million barrels—that's 395,000 barrels of oil per day (nearly equal to the daily energy consumption of Greece).

For just the first three weeks of combat in Iraq, the Army calculated that it would require more than 40 million gallons of fuel—an amount equivalent to the total gasoline used by all Allied Forces combined during the four years of World War I.

For fiscal year 2004, the DESC spent its yearly budget of $3.5 billion for 110 million barrels of petroleum products. Or, the DESC boasts: "That's enough fuel for 1,000 cars to drive around the world 4,620 times—or 115.5 trillion miles."

Military vehicles have no respect for Corporate Average Fuel Economy standards—or fuel standards of any kind. After all, we are at war; fuel economy is a luxury. Indeed, the military refers to fuel consumption in terms of "gallons per mile," "gallons per minute," and "barrels per hour." One quickly realizes that military "assets," as the Pentagon likes to call its rolling arsenal, operate in a world all their own, free of restraints of any kind—both in the fuel they consume and the pollutants they exhaust. (The military is now using more highly toxic JP-8 jet fuel for many of its vehicles.)

The stalwart of all the fighting vehicles, the M-1 Abrams tank, according to the military's own spec sheets, gets 0.2 miles per gallon—that is, it needs five gallons of fuel to cover a single mile. Just firing up the tank's turbine requires ten gallons of fuel. During battle, over ideal terrain, each of the Army's 1,838 Abrams tanks can go through 300 gallons—usually JP-8 jet fuel—every hour.

Feeding the appetites of these voracious machines requires intricate logistical planning and support from some 2,000 trucks, a battery of computers, 20,000 GIs, and as many as 180,000 workers under federal contracts—more contract workers than soldiers.

In 2002, Gen. Paul Kern, head of the Army Materiel Command, revealed that the actual cost of fuel (depending on how it gets delivered) can range anywhere from a low of $1 to a high of $400 per gallon. The average, he allowed, hovered around $300 a gallon.

Floating and Flying Gas-Guzzlers

At its top speed of 25 knots per hour, the now-decommissioned aircraft carrier *USS Independence* consumed 134 barrels of fuel an hour, or close to 5,600 gallons an hour. (The ship boasts 4.1 acres of flight deck and a crew of 2,300.) On its way to the Persian Gulf in 1991, a trip that took fourteen days, the *Independence* went through two million gallons of fuel. Every four days, the ship took on an additional one million gallons of fuel, half of which went to supply the carrier's ninety fighter jets.

According to the 2006 *Navy Almanac,* at the beginning of 2006, the Navy held an inventory of 285 combat and support ships, along with 4,000 planes and helicopters. (The number and kinds of vessels stationed in the Persian Gulf was classified.)

Of all the branches, the Air Force uses the most fuel. In 2006, the Air Force consumed nearly half of the DoD supply—2.6 billion gallons of jet fuel, the same amount of fuel consumed from December 1941 to August 1945, during World War II. The Apache helicopter blows through fuel at an astonishing rate. Powered by two General Electric gas-turbine engines, each rated at 1,890 horsepower, the Apache gets about one-half mile to the gallon. One pair of Apache helicopter battalions in a single night's raid will consume about 60,000 gallons of jet fuel. Any of the large helicopters—the Sea Stallion, Super Stallion, Sea Dragon, or Pave Low III—sucks up five gallons every mile. But that's nothing compared with the fighter planes.

With its afterburners fired up, the F-16 fighter jet uses 800 gallons per hour and the F-15 about 1,580 gallons per hour. More dramatically, the F-4 Phantom Fighter uses 40 barrels of fuel (more than 1,600 gallons an hour). But the gas-hog award goes to the B-52 Stratocruiser, which has eight jet engines and zips through an astonishing 86 barrels of fuel (roughly 3,334 gallons) every hour. In one hour of flight—600 miles—the B-52 uses as much fuel as the average car driver uses in seven years.

Even though the B-52H holds an enormous 47,975 gallons of fuel, it still requires mid-air refueling. The KC-10 aerial refueling tanker burns 2,050 gallons per hour while the larger KC-135 Stratotanker (which carries 31,275 gallons of fuel) sucks up an impressive 2,650 gallons per hour. The amount of fuel that the KC-135 carries would keep the average family car running for 62.5 years.

The Pentagon makes public very few statistics about the B-52 and F-117 Stealth fighters, except that the B-52 can carry sixteen 2,000-pound laser-guided bombs or eighty 500-pound laser-guided bombs. It is remarkable that we even know that the stealth fighters exist.

Pentagon Pollution

Given that the DoD is the world's largest single consumer of fuel, it is no surprise that the Pentagon also is the planet's largest polluter. The DoD provides little or no data for its many thousands of vehicles. While the great majority of vehicles burn gasoline, others use diesel, and still others jet fuel. The Abrams Tank can burn whatever is available, making it even more difficult to determine exact CO_2 numbers. I base my calculations on the Environmental Protection Agency's fact sheet EPA420-F-05-004 on greenhouse gas emissions from passenger vehicles.

Thus, if we simply use the DESC's own figures for 2004, those 4,620 trips around the world would consume approximately 2.2 billion gallons of fuel, sending a shocking 24 billion tons of carbon dioxide, a greenhouse gas, into the atmosphere. No one outside the Pentagon can say with certainty just how much fuel the Navy actually uses. It may even be higher than the Air Force.

The Pentagon places the fuel it reserves for supposed international purposes in a category called International Bunker Fuel. But Bunker Fuel (or "Bunker Oil") remains off the record, as nonexistent as the prisoners held at Guantanamo Bay. Bunker Oil contains higher concentrations of sulfur than other diesel fuels, and so pollutes not just with CO_2 but with SO_2 (sulfur dioxide) as well. The two gases in combination do more damage to the environment, for they form a thicker layer in the atmosphere and capture heat more tenaciously. The United States does not figure into its own annual CO_2 numbers any of the greenhouse gases that the military generates.

Some environmentalists insist that aircraft carriers pollute more than any other piece of armament in the military arsenal. Besides spreading the ocean surface with its own CO_2 and residual oil, seagoing vessels create "ship tracks" that tail off like vapor trails in the atmosphere and have the potential for changing the microstructure of marine stratiform clouds. Made up of sulfur dioxide, nitrous oxide, and water molecules from both diesel-powered and steam-turbine powered ships, these long-lived clouds, according to some studies, help to intensify the greenhouse effect.

Along with emitting CO_2, jet planes pump out a trail of nitrous oxide and sulfur and water particles, which, according to some toxicologists, may be three times more harmful to the atmosphere than CO_2 alone. Kerosene jet fuel puts considerably more carbon per gallon into the atmosphere than gasoline or diesel. At altitude, airplanes spew not only carbon dioxide but also nitrous oxide, sulfur dioxide, soot, and water vapor—a combination that may triple their total warming effect on the climate. Supersonic aircraft, like the Super Hornet, the F-111, and the F-22 Raptor, create pollution 5.4 times more corrosive to the environment than conventional commercial aircraft.

Trying to calculate CO_2 pollution for military flying is near impossible. For one thing, we know nothing of the stealth F-117's fuel consumption. We do know, however, that sorties for that plane at the beginning of the Iraq War lasted 1.6 hours. Flying out of distant bases raised the average sortie time to 5.4 hours, with some sorties lasting up to 7 hours—refueling accomplished in the air. Forty-two F-117s each flew more than 1,300 combat sorties. The total amount of carbon dioxide that went into the atmosphere remains unreported.

A March 2007 draft report prepared by the George W. Bush administration for the United Nations estimated that U.S. emissions of greenhouse gases would rise from 7.7 billion tons in 2000 to 9.2 billion tons in 2020—an increase of 19.5 percent. But the report was entirely silent on the amount of greenhouse gases that the military was pumping into the atmosphere—twenty-four hours a day, every day, during the long years of war in Iraq and Afghanistan.

Superpower, Superpolluter

Tyrone Savage, War Resisters League (2000)

Thanks to the U.S. military's relentless abuse of the environment, the world's sole remaining superpower is also its leading superpolluter.

Failing to maximize the opportunities opened up by the end of the Cold War, the United States continues to ravage the world ecosystem by creating, testing, and deploying enough deadly weaponry to destroy any enemy many times over (while spending more on its military than the combined total spent by "rogue" nations Iran, Iraq, Libya, Syria, Cuba, Sudan, and North Korea). The environmental impact of the Pentagon's activities is substantial:

- The U.S. military produces nearly a ton of toxic pollution a minute. The GAO estimates that the Department of Defense generates 500,000 tons of toxics annually—more than the five leading chemical companies combined.
- The Pentagon reported to Congress in 1990 that more than 17,484 military sites violated federal environmental laws. At least ninety-seven were on the Superfund list, which designates a cleanup to be a national priority.
- U.S. military activities—excluding the manufacture of weapons—annually consume enough energy to run the entire U.S. urban mass transit system for twenty-two years.
- Forty thousand underground tanks used to store military chemicals and fuels threaten nearby communities. In New Jersey, 3.2 million gallons of aviation fuel and other chemicals leaked from the Lakehurst Naval Air Station and contaminated an aquifer that provides tap water for half of the state. Tests found toxic levels 10,000 higher than the government considers safe.

In 2014, the United States budgeted $526.6 billion for global military spending—approximately $1 million per minute. Washington spends sixty-six times more money for the Pentagon than it assigns to the Environmental Protection Administration. Two days' worth of military spending could halt desertification of the world's threatened areas and just five minutes' worth could protect endangered species and combat ocean pollution for one year.

The environmental impact of U.S.-led global militarization is not only seen in the "peacetime" consumption of resources within our borders; it is also conspicuously reflected in the foreign outposts of U.S. influence—at overseas military bases and, even more appallingly, in U.S. wars waged in foreign lands.

The Pentagon's policy for domestic military sites features early identification of toxic sites and calls for effective, cost-efficient cleanup. But after overseas bases close, the Pentagon's pollution problems are dumped on the host countries—often without the documentation needed for a cleanup. In the Philippines, after the Pentagon's 1992 evacuation of Subic Naval Station and Clark Air Base, residents found that tons of toxic chemicals had been dumped on the ground and in the water or buried in uncontrolled landfills. At U.S. bases in Germany, industrial solvents and other hazardous waste have destroyed nearby ecosystems. Subsequent restoration efforts were covered by German taxpayers.

On the Puerto Rican island of Vieques, which the U.S. Navy used as a bombing range for half a century, the U.S. Navy fired radioactive depleted uranium munitions into the Vieques training range—in blatant violation of both Nuclear Regulatory Commission rules and the Navy's own official training regulations. (Only 57 of 263 DU shells fired into the island's waters were recovered.) Cancer rates on Vieques are 28 percent greater than on the main island.

In Panama, U.S.-run artillery ranges are littered with more than 120,000 pieces of unexploded ordnance. The United States conducted an active chemical weapons program in Panama for more than forty years, and chemical warfare agents remain buried in the Panama Canal Zone. Department of Defense cleanup plans would leave 8,000 acres at three artillery ranges untouched, despite the proximity of the 60,000 people living in adjacent communities.

An overseas cleanup policy promulgated by the Department of Defense in October 1995 was vastly weaker than domestic law and profoundly unfair to host countries. The Pentagon was only obligated to remedy hazards that it deemed "imminent and substantial." Additional cleanup was required only if doing so helped military operations, was required by international health and safety agreements, or was funded by the host country.

During the 1990s, U.S. expenditures on overseas base cleanups averaged approximately 1 percent of that spent on domestic base cleanups. Host countries often lack the financial or technical resources to clean up after the U.S. military leaves. This adds insult to injury and violates the generally accepted principle of "polluter pays."

Efforts to reduce the U.S. stockpile of nuclear warheads by dismantling these weapons has led to a challenging problem: what to do with the radioactive leftovers, including highly enriched uranium and plutonium pits with a half-life of 24,000 years. Depleted uranium, the radioactive waste left over when the isotope uranium-235 is extracted from naturally occurring uranium, can deliver a lethal exposure in seconds to someone standing unshielded three feet away. The United States has more than one billion pounds of DU waste—and no safe method of disposing of it. Instead, it sells the DU to industries that produce super-dense antitank shells that have been used in several U.S. wars over the past decades.

During the Gulf War, a significant portion of the 630,000 pounds of armor-piercing depleted uranium rounds fired inside Iraq and Kuwait left behind enough DU to cause tens of thousands of cancer deaths.

During the bombing of Yugoslavia, the U.S. Joint Chiefs of Staff confirmed that DU was used in the shell casings fired against Serbian forces by A-10 Warthog jets and in the nose cones in Tomahawk missiles. The British newspaper *The Guardian* called the bombing "the dirtiest war the West has ever fought. . . . A war which targets chemical factories and oil installations, which deploys radioactive weapons in towns and cities, is a war against everyone: civilians as well as combatants, the unborn as well as the living."

While the drastic environmental impact of the U.S. military's activities around the world remains relatively unknown inside the United States, other nations are painfully aware of it. Delegates to the 1992 Rio Earth Summit went so far as to demand that the agenda include the impact of militarism on the environment and on development.

The United States vetoed that agenda item.

Jet Fright: The Impacts of Military Aircraft

Petra Loesch (1989)

Military aircraft are both consumers and polluters. It takes nearly 4.5 tons of titanium, nickel, chromium, cobalt, and aluminum to manufacture a single F-15 jet engine. Flying miles above the Earth, these engines burn through tanks of JP-8 kerosene and spill global-warming exhaust trails of water vapor and chemicals—including nitrogen oxides that freeze into ice clouds, which in turn accelerate the destruction of the planet's protective ozone shield. But the impact of these powerful weapons is felt closer to the Earth as well.

On June 14, 1989, after four years of heated debate, the Canadian government gave the United States permission to fly low-level bomber training missions over the Canadian Northwest. This decision was vigorously opposed by Canadian environmental groups and Indigenous peoples who feared the jets would disturb wildlife and disrupt native traditions. As it turned out, these concerns were well founded.

There was no lack of information about the human health impacts of powerful jet engines screaming over rural rooftops. Much of this information came from Germany, where the United States routinely launched military flights from two sprawling air bases.

"I had to cover my children's ears because I was so afraid that they could be injured," one German mother recalled. "My son was so scared that I had to take him to the hospital . . . his face was pale with fear." The young mother was talking about the terror of lower-level military jet flights over Germany, but the location doesn't matter—it could be Canada, Turkey, or the American Southwest—for wherever fighter jets conduct their flyovers, the impacts on human health, wildlife, and wilderness can be environmentally devastating and, frequently, deadly.

During the 1960s to 1980s, West Germany experienced nearly 600,000 military flights per year, 100,000 of which were lower-level flights ranging as close as 225 feet above the ground. This densely populated country was one of the favorite playgrounds of the German air force and its NATO allies—the United States, Great Britain, and Canada. Most of NATO's air crashes occurred in West Germany.

In 1988 alone, almost 120 people died as a result of military crashes in West Germany, and U.S. pilots were involved in the majority of these accidents. In 1988, the rate of U.S. crashes doubled—owing, in part, to two catastrophes: one at the Ramstein Air Force Base (where seventy spectators died during an air show collision) and another in the town of Remscheid (where a U.S. A-10 Thunderbolt crashed on a street, killing six people and setting several houses on fire).

Aside from these major crashes, casualties also resulted from "small losses"—incidents where fighter jets lost parts of their equipment (rockets, bombs, external flight recorders, kerosene-filled tanks) in or near residential areas, watersheds, or wildlife sanctuaries. (In one case, the external flight recorder from a Tornado jet fell into a beer garden in Munich restaurant.)

Even the air blast caused by low-flying jets can prove devastating. In the Bavarian city of Bad Tölz, the roof of a baroque church caved in after a jetfighter roared by a bit too close. In other cities, roofs of houses have been torn off, roofing tiles have been sent flying, and cathedrals have sustained damage.

The military's "lowest-level flight zone" allows flights over nonresidential areas to drop as low as ninety-two feet above the ground "to demonstrate bacteriological and chemical war strategy." Following a growing storm of complaints about flights below the minimum level and violations of night-flight prohibitions, Germany's defense ministry gave suffering citizens a break—by prohibiting flights during the lunch hour. But even this regulation was violated.

Kerosene Rain

Environmental destruction caused by military air traffic is more serious than officially reported, and often underestimated. As the amount of jet fuel used continues to increase, more and more pollution rains to earth as an invisible drizzle, bringing slow death to forests, lakes, and fields. Experts warn that a single liter of kerosene can contaminate a million liters [264,172 gallons] of groundwater.

During every hour of flight, a jet burns nearly 8,000 gallons of kerosene. Five percent of this is blown into the air as carbon monoxide, nitric oxides, hydrocarbons, soot, and carbon dioxide—up to 10 million kg [11,000 short tons] a year, mostly concentrated in lower-level flight areas.

In addition to burning immense amounts of fuel, jet engines also consume vast amounts of oxygen. A passenger jet can gulp down 528,344 gallons of air per second during takeoff and, in the first five minutes of flight, burn as much oxygen as a 44,000-acre forest can produce in a day.

During each routine refueling procedure (from fuel truck to storage tank to tank-vehicle to jet), approximately eight gallons of kerosene are lost. When a fighter jet prepares to land, safety rules require that pilots jettison excess fuel through exhaust valves. (In an "emergency," all excess fuel can be dumped into the air.) This routine fuel dumping in the vicinity of airfields has sparked fires near U.S. bases.

Kerosene is also spilled over land, lakes, and oceans when tanker jets transfer fuel to jet fighters during mid-air refueling maneuvers. These exercises are supposed to be conducted away from residential areas at altitudes of 1,600 to 2,900 feet, but citizens' complaints (and photographs) prove that this is not always the case.

Runoff on the Runways

Even when parked on the ground, military aircraft can pose a pollution problem. Fighter jets need to be repainted in camouflage colors on a regular basis. During this repainting procedure, crews don safety suits and masks to protect themselves from toxic fumes. The process of removing the old paint creates an even greater problem as chemical solvents and paint chips laden with heavy metals are washed away into the soil.

To keep airports ice-free during winter, the military uses de-icing compounds, which consist mainly of urea. The runoff can rob rivers and lakes of oxygen and can lead to nitrate contamination of the groundwater. At the U.S. base at Bitburg, Germany, several mass fish kills have occurred in the nearby Kallenbach River. One investigation found 163 milligrams (mg) of urea per liter of water. One mg/L can be toxic to fish.

F-16 jets flown by the United States, Belgium, The Netherlands, and Denmark, hold another hidden danger: hydrazine, a toxic liquid that serves as an "emergency power unit" for restarting stalled engines.

Hydrazine is a clear, ammonia-scented liquid that forms explosive, heavier-than-air mixtures. On hot days, spilled hydrazine vapors can "crawl" invisibly over the ground. When ignited, the explosions can ricochet over long distances. Hydrazine vapors irritate eyes and respiratory organs and contact can burn the skin. Absorbed through the skin, hydrazine can damage the liver, the central nervous system, the heart, kidneys, and blood. The lethal dose is one gram. Every F-16 carries nearly seven gallons of hydrazine.

Jet Noise and Health Effects

Jet noise is extremely intense—a visceral assault the U.S. Air Force likes to call the "Sound of Freedom." During an average flight at 246 feet above the ground, noise levels of up to 125 decibels (dB) are quite normal. In comparison, a jackhammer produces between 90 and 100 dB. Low-level flight can be more than four times louder than the loudest jackhammer, due to the fact that every ten-decibel increase represents a *doubling* of the noise level.

Because a jet's "noise-burst" comes fast and without any warning, the body has no way to prepare itself for the shock. A typical "alarm reaction" to a low-level noise-burst triggers an adrenaline release that can last for about four hours, causing high blood pressure, increased heart rates, and disturbances of the intestinal tract and other organs.

Children seem to be the most affected. Youngsters frightened by jet noise become panic-stricken, succumb to crying, and cannot be comforted. Many children exposed to military jet noise have been found with threshold hearing levels impaired by as much as 30 percent. Some children refuse to go outside on sunny days knowing that it is ideal weather for fighter jets. Other children suffer from insomnia and wake up at night screaming.

A study by the *Bundeswehr* (Germany's federal defense forces) revealed that when fighter jets screamed by overhead, grazing cattle moved closer together; horses trembled nervously and began to bite and kick; dogs showed increased heart rates, began to bark, ran in circles, and tried to hide; zoo animals got diarrhea. The study also reported spontaneous abortions and shortened gestation in cattle. Low-level overflights in Canada triggered caribou stampedes and caused terrified foxes and minks to cower. Nesting birds have abandoned their nests (long enough for local predators to dine on their eggs), hens stopped laying eggs or began to devour them. Sometimes eggs are destroyed in the nests by the 140 dB shockwaves of jets passing over the treetops.

Vogelschlag Collisions

Pilots favor tidal marshlands because their low, flat stretches allow jets to drop down to 246 feet and fly over long distances. But these tidal flats are an important breeding, molting, and resting habitat for many types of birds, especially for migrating waterfowl who flock to the marshes to rest and gather the necessary fat reserves needed for their winter migrations. Because migratory birds usually rest in large groups, the alarm reaction of one bird causes all the other birds to fly up. Birds disturbed by rocketing aircraft may not have enough time to feed, which may affect their ability to successfully complete their migrations.

Another danger is *Vogelschlag* ("bird strikes"), caused when single birds or flocks collide with jets, often resulting in a deadly crash. *Vogelschlag* incidents are most likely to occur during the great southern migrations. The most numerous victims are seagulls, doves, raptors, and marshland birds.

Almost 80 percent of the jet crashes, however, are due to human failure. The most dangerous situations occur when the pilots pull up their jets really fast. Crushed by pressures of more than sixteen times the force of gravity, vision decreases to zero, causing a visual "gray out," a narrowing field of vision and, finally, a "blackout" and loss of consciousness for up to twenty-five seconds. At speeds topping 500 miles per hour, a jet can cover well over four miles before a pilot fully regains consciousness.

NATO Planes in North American Skies

In response to increasing citizen complaints, Germany's air force relocated their aerial maneuvers to Canada and Turkey. During the 1980s and 1990s, thousands of hours of "educational" exercises were conducted in the skies around the Canadian Forces Base at Goose Bay, Labrador. NATO aircraft from Germany, Britain, The Netherlands, and Italy roared overhead at speeds of more than 500 mph, sometimes just 98 feet above the ground.

Canada set aside more than 38,610 square miles of airspace over the Quebec-Labrador Peninsula for military training, conducting upwards of 30,000 sorties a year (a third of these flown at night). The Cold Lake Air Weapons Range claims 173,746 square miles of airspace over Saskatchewan and Alberta and was recognized as one of the world's largest "live-drop training ranges" and North America's largest low-level flying zone.

In the United States, California's Mojave Air and Space Port hosted as many as 90,000 training flights a year, crisscrossing 18,147 square miles of desert. *Worldwatch* has estimated the average annual global total of military sorties at between 700,000 and a million.

In North America, fourteen ancient native communities were placed under the constant sonic barrage of military jets. Labrador (known as "Nitassinan" to its native peoples) is the land of the Innu—one of North America's last remaining hunting cultures—and home to vast herds of caribou. In 1972, the Innu were relocated to a reservation and the Canadian government declared the region officially "unpopulated."

Nonetheless, the Innu still continue to migrate over the land throughout the year. Protesting the "military invasion of our homeland," Innu chief Daniel Ashini complained: "NATO jets fly down river valleys, over lakes and marshes. These are the best areas for hunting, trapping, fishing. . . . They have turned our hunting grounds into a wasteland for war games."

Land Mines: The Smallest WMDs

Jody Williams, International Campaign to Ban Landmines (1997)

> *What weapon is still lethal to unsuspecting human targets when the soldiers who brought it to the battlefield have become old men?*
>
> —Russell W. Ramsey, U.S. Army School of the Americas, Fort Benning, Georgia

The history of land mines can be traced back to the American Civil War. But mines as they are known today were originally developed during the First World War to defend against tanks. Given the size of antitank mines, it was relatively easy for enemy troops to enter minefields and remove the weapons for their own use. This led to the development of the antipersonnel mine, a much smaller delayed-action explosive device that was sown throughout antitank minefields to deter enemy soldiers from entering. First used to protect the more valuable antitank mine, the anti-personnel mine has taken on a life of its own.

Although they were originally designed for use primarily as defensive weapons, land mines have increasingly been deployed offensively. While such use has not been confined to internal conflicts—the United States pioneered advances in mine technology and deployment during the war in Indochina, and the former Soviet Union resorted to them on a massive scale in Afghanistan—land mines have become a choice weapon in these wars and their offensive use often is a preferred tactic. Cheap, easily available, and "ever vigilant" once emplaced, antipersonnel land mines have proliferated in armed conflicts everywhere.

According to the United Nations, "every year, landmines kill 15,000 to 20,000 people—most of them children, women, and the elderly—and severely maim countless more."

What sets the land mine apart is its time-delay function. Not designed for immediate effect, land mines lie dormant until triggered by a victim. In many cases, land mines have been used as offensive weapons to cut off access by opposition forces and their civilian supporters to large tracts of land.

Often designed to maim, their psychological impact on the enemy is undeniable. In addition to demoralizing combatants, land mine casualties can also overload military logistical support systems since most mine victims require more extensive medical and rehabilitative attention than other types of war-related casualties. Moreover, land mines do not discriminate between the logistical support systems of the military and those of society as a whole. They terrorize and demoralize civilians, and their impact on fragile, local health systems can be overwhelming. Post-conflict land mine casualties are almost exclusively civilian.

Unlike other weapons of war, landmines and explosive devices, that act like landmines, were not silenced by any peace agreement. That changed with the 1999 ratification of the Ottawa Treaty, which prohibited the use, stockpiling, production, and transfer of land mines. As of 2017, 162 countries were parties to the treaty. So far, the United States (along with China and Russia) has refused to sign the agreement.

Nature and Scope of the Problem

Land mines have been used on a massive scale since their development. In 1995, it was estimated that 400 million land mines had been sown since the beginning of the Second World War, including at least 65 million added since 1980. An estimated 80 to 110 million land mines have been deployed in at least 64 countries around the world. The majority of countries most heavily contaminated with land mines are in the developing world.

Africa is the hardest hit continent, with a total of perhaps 37 million land mines in at least 19 countries. Angola alone has an estimated 10 million land mines and an amputee population of 70,000. Other countries particularly affected are Eritrea, Ethiopia, Mozambique, Somalia, and Sudan. But Africa is not alone—mines are also found in Asia, Europe, Latin America, and the Middle East.

While land mines are ubiquitous, they have been used in particularly large numbers in Afghanistan, Angola, and Cambodia. There are at least 28 million land mines in those three countries alone, which are home to 85 percent of the world's land mine casualties. Europe is said to have the fastest-growing problem, with more than three million land mines deployed during the fighting in the former Yugoslavia.

Land mines have been used so extensively because they are readily available, cheap, and easy to use. While land mines are not hard to deploy, their removal is painstakingly slow, dangerous, and expensive. Mine-detection technology has not kept pace with rapid developments in mines, which have made them more deadly and more difficult to trace.

Mines, which used to be made of metal and thus were relatively easy to find, are now increasingly made of plastic. Currently available systems do not reliably detect minimum-metal plastic mines in battle-contaminated field conditions. In Cambodia, for example, for every mine found, an average of 129 harmless metal fragments are detected. . . . Mines have become sophisticated weapons with electronic fuses and sensor systems, which can make them even more deadly. They can now sense footstep patterns, body heat, sound, and the signal of a mine detector—all or any of which can make them explode.

Clearance is made even more difficult by an almost complete disregard for the stipulated mapping and recording of minefields. While the 1980 Convention on Conventional Weapons requires the mapping of "preplanned" minefields, the term "preplanned" is not defined. Even if it were, given the few instances of minefields mapping and recording in the majority of conflicts of the past several decades, the provision would probably not be followed. Military instructions also provide for the mapping and recording of minefields. But as the United Nations and experts involved in humanitarian mine clearance have repeatedly pointed out, in the overwhelming majority of cases, instructions in this regard are not heeded.

Advances in mine-delivery systems have made it possible to remotely scatter mines at rates of well over 1,000 per minute. While it might be possible to record the general location of such mines, even the military concedes that accurate mapping is impossible.

Even in the relatively few instances where minefields have been mapped, in many cases the information has become almost irrelevant over time, as weather conditions have changed the original location of the weapons. For example, mines sown on riverbanks have been washed downstream by flooding, and mines sown in desert environments move easily and frequently in shifting sands. Also, in heavily contested areas, mines are often sown on top of previous minefields so that even if maps have been made at some point during the conflict, they do not include all of the new mines laid as battlefronts shift back and forth and opposing forces mine and re-mine the same areas.

Humanitarian mine clearance involves the removal of all mines (the UN standard is 99.9 percent) to return previously mined land to civilian use. Numbers alone do not fully explain the problem. It takes 100 times as long to remove a land mine as to deploy it. And a field with one land mine in it can be unfit for productive use as surely as a field with 100 land mines. It can take a mine-removal team as long to clear a field with one mine in it as a field with 100 mines.

The process is the same wherever there is a fear of mine contamination: the entire area must be painstakingly combed and probed to remove mines that are actually there—or to demonstrate that the area is free of mines. With the millions of land mines currently contaminating the globe, even if no more mines were produced or deployed, it would take decades to overcome the problem.

Of the more than 255 million land mines manufactured over the past 25 years, about 190 million have been antipersonnel mines. At one time or another, at least 100 companies were involved in the production of 360 types of antipersonnel mines in 55 countries. Current production averages about 5 million mines every year; for the previous 25 years, it averaged 10 million annually. Of the $20 billion spent annually on arms, it is estimated that conventional antipersonnel mines account for less than $100 million.

China, Russia, and Italy have been the major producers and traders of land mines in recent years. Other important suppliers have included the former Czechoslovakia and the former Yugoslavia, along with Egypt, Pakistan, and South Africa. Prior to the mid-1980s, the United Kingdom, Belgium, and the United States ranked among the top producers and exporters; other significant exporters in that period included Bulgaria, France, and Hungary.

The Campaign to Ban Land Mines

Clearing land mines while others were being planted, manufactured, and traded was no solution. Amputating limbs and providing prostheses for one survivor while another bled to death unaided was no solution. Why provide improved seeds for farmers whose fields were mined, or vaccinate animals that graze in minefields? We saw a world where peace had few advantages over war. The circle of manufacture, supply, and use had to be broken. The answer was a ban—and so the International Campaign to Ban Landmines (ICBL) was born.

The ICBL has grown to include approximately 350 NGOs working in more than 20 countries around the world for a ban on antipersonnel land mines. This is a twofold call for (1) an international ban on the use, production, stockpiling and sale, transfer, or export of antipersonnel mines and (2) for contributions, by countries responsible for the production and dissemination of antipersonnel mines, to the international fund administered by the United Nations and to other programs to promote and finance mine-victim assistance and land mine awareness, clearance, and eradication worldwide.

And so we view the 1997 Ottawa Treaty as a first and valuable step, a milestone in a battle to rid this world of antipersonnel mines. While these weapons remain in the world's armories, there is no nation immune from their effects.

What of those nations that have failed to sign the Ottawa Treaty? It would be easy to focus totally on China, the United States, and Russia, nations whose stubborn refusal to put humanitarian concern above ill-judged military policy is inconsistent with their status as UN Security Council members and major regional powers.

But what of those countries like South and North Korea, India, Pakistan, Israel, and Syria, whose often-valid concern for their border defenses blinds them to the damaging nature of the antipersonnel mine? What of Egypt, a country blighted by land mines emplaced decades ago, which argues it needs antipersonnel mines to deter smugglers? South Korea and the United States argue that the Demilitarized Zone minefields are of such importance they wish to make them exempt from any land mine ban.

Freedom is so often the justification for war. But where is the sense in fighting for the freedom of a people employing a weapon which will deny those same people, in peacetime, freedom to live without fear, freedom to farm their land, freedom merely to walk in safety from place to place—deny them the freedom to let their children play without being torn apart by a land mine? That is no freedom.

All those states that have failed to sign this treaty have failed humanity. They are intransigent and uncaring in the face of compelling humanitarian, economic, and environmental evidence that antipersonnel mines should be banned.

Antipersonnel mines do not only sever limbs; they can break the human spirit. In most mine-affected countries we, the international community, must offer more than the surgeon's knife and prostheses as support to those who survive the blast of a land mine. In the same way as the Ottawa Treaty is only the first step toward a global ban, so prostheses should be seen as the first stage in the support process for the victim of a mine blast. The reason for this lack of response is evident and shames us all—we simply do not care enough. This is a responsibility that the ICBL places high on its action agenda. We must have respect for the rights of those who fall victim to land mines, most importantly their right to control their own lives and their right to be heard.

There is a popular myth that mine clearance costs too much—the ICBL does not accept that is true and, faced with the obscenity of the effects of the antipersonnel mines, it would be difficult to understand what scale of measurement could be used to make such a calculation. We *must* afford it; we cannot talk of having concern for the global environment and yet leave future generations a blighted world with land made unusable by this deadly military garbage. The tens of millions of dollars spent annually on mine clearance pale in comparison to the hundreds of billions spent on the military. Military organizations—the parties primarily responsible for the laying of land mines—are polluters who can afford to pay the price of clearance.

Explosive Arsenals: "Bomblets" to "Near-Nukes"

Helen Caldicott, International Physicians for the Prevention of Nuclear War (2017)

The 1991 Persian Gulf War was, in effect, a nuclear war. By the end of Operation Desert Storm, the United States left between 300 and 800 tons of depleted uranium-238 in antitank shells and other munitions scattered across the battlefields of Iraq, Kuwait, and Saudi Arabia.

The term "depleted" refers to the removal of the fissionable element uranium-235 through a process that ironically is called "enrichment." What remains, uranium-238, is 1.7 times more dense than lead. When U-238 is incorporated into an antitank shell and fired, it achieves enormous momentum, cutting through tank armor like a hot knife through butter.

U-238 is pyrophoric. When it hits a tank, it bursts into flames, producing aerosolized particles less than five microns in diameter, making them easy to inhale. U-238 is a potent radioactive carcinogen that emits a relatively heavy alpha particle composed of two protons and two neutrons. Once inside the body—either in the lung if it has been inhaled, in a wound if it penetrates flesh, or ingested (since it concentrates in the food chain and contaminates water)—it can produce cancer in the lungs, bones, blood, or kidneys.

U-238 has a half-life of 4.5 billion years, meaning the areas in which this ammunition impacted in Iraq and Kuwait will remain effectively radioactive for the rest of time.

Children are ten to twenty times more sensitive to the effects of radiation than adults. My fellow pediatricians in the Iraqi city of Basra have reported an increase of six to twelve times in the incidence of childhood leukemia and cancer. The incidence of congenital malformations doubled in the exposed populations where these weapons were used. Babies were born with horrific tumors, without eyes, without brains, crippled with stunted and deformed limbs, internal organs positioned outside the body, fingers and toes fused together.

And the medical consequences of U-238 exposure was not limited to Iraqis. Some exposed American veterans were reported to be excreting uranium in their urine a decade later. Nearly one-third of the American tanks used in Desert Storm were armed with munitions made with U-238, which exposed their crews to extensive whole-body gamma radiation. The Pentagon's own studies prior to Desert Storm warned that aerosol uranium exposure under battlefield conditions could lead to cancers of the lung and bone, kidney damage, non-malignant lung disease, neurocognitive disorders, chromosomal damage, and birth defects.

"Enduring Freedom," Enduring Devastation

During the first four weeks of Operation Enduring Freedom, which began in October 2001, the United States dropped half a million tons of bombs on Afghanistan—twenty kilos (forty-four pounds) for every man, woman, and child. During eight-and-a-half weeks of U.S. bombing, a documented 3,763 civilians were killed. Some of the conventional weapons America used to support the Northern Alliance during their advances on the Taliban were so powerful that they are described by the Pentagon as "near nuclear" weapons. Among America's arsenal of "conventional" ordnance are the following:

Cluster Bombs: Terrifying and deadly, each cluster bomb is composed of 202 bomblets packed with razor-sharp shrapnel designed to be dispersed at super-high speed, ripping into human bodies over an area the size of 22 football fields. These weapons are prohibited by the Geneva

Protocol. Civilians were inevitably killed throughout Afghanistan by these illegal and dreadful weapons. On one documented occasion, the United States bombed a mosque in Jalalabad during prayer and, while neighbors were digging out 17 victims, additional bombs killed more than 120 people.

Historically, between 5 percent and 30 percent of these bomblets fail to explode initially, lying around the countryside as mines that detonate with violent force if touched, tearing their victims to pieces. Tragically, the bomblets are colored yellow and shaped like a can of soft drink, and therefore attractive to children. (Food parcels containing peanut butter, Pop Tarts, rice, and potatoes dropped throughout Afghanistan by the United States are also yellow and the same size and shape as the munitions.) Human Rights Watch estimates that over 5,000 unexploded cluster bomblets may be littered across Afghanistan, adding to the hundreds of thousands of mines left after the 1979–89 Soviet-Afghan war. Afghanistan is currently the most heavily mined country in the world.

Bunker Busters: Dropped from B-1 or B-2 planes, these 5,000-pound behemoths are made from the gun barrels of retired naval ships and are so heavy that they burrow 20 to 100 feet into the ground before their high-explosive materials detonate. Most are laser guided, but some use global positioning satellites for guidance.

"Near-Nukes": The Foreign Military Studies Office at Fort Leavenworth claims "a fuel air explosive can have the effect of a tactical nuclear weapon without the radiation." There are many different varieties of "thermobaric" Fuel-Air Explosives (FAEs), but they typically consist of a container of fuel and two separate explosive charges. Dropped by parachute from a 78-ton Lockheed MC-130, they detonate just above the ground, creating a wide area of destruction.

The first explosion bursts the container at a predetermined height, disbursing the fuel, which mixes with atmospheric oxygen. The second charge then detonates this fuel-air cloud, creating a massive blast that kills people and destroys unreinforced buildings. Near the ignition point, people are obliterated, crushed to death with overpressures of 427 pounds per square inch and incinerated at temperatures of 2,500 to 3,000°C. Another wave of low pressure—a vacuum effect—then ensues. People caught in this second zone of destruction are severely burned and suffer massive internal organ injuries before they die. In the third zone, eyes are extruded from their orbits, lungs and eardrums ruptured, and severe concussion ensues. The fuel itself—ethylene oxide and propylene oxide—is highly toxic.

On April 13, 2017, the Pentagon dropped a GBU-43/B Massive Ordnance Air Blast (MOAB) bomb on Nangarhar Province, Afghanistan. The 21,000-pound fuel-air device exploded above the targeted area with the force of 11 tons of TNT, causing massive damage. Moscow quickly let it be known that Russia has devised a similar "near-nuke" weapon. Russia's Aviation Thermobaric Bomb of Increased Power is smaller and four times as powerful as the Pentagon's bomb. (The US nuclear bomb that destroyed Hiroshima had an explosive yield of 15,000 tons of TNT.)

The Nuclear Option: The United States' behavior in Afghanistan has veered frighteningly close to deployment of nuclear weapons. In addition to deploying the most horrific conventional weapons, the Defense Department has recommended the use of tactical nuclear weapons, while some members of Congress have advised the use of small nuclear "bunker busters." A number of George W. Bush's advisors also advocated the use of nuclear weapons, while neutron bomb inventor Samuel Cohen postulated that his weapon might be appropriate for Afghanistan. (Neutron bombs have a relatively small blast effect and can expose large numbers of people to deadly radiation impacts while leaving buildings intact.)

On September 11, 2001, before much of the world was even aware of what happened in New York, Washington, and Pennsylvania, the Bush administration had raised the country's nuclear alert codes from DEFCON 6 to DEFCON 2—the highest state of alert before the launch code is operable. Russia (the country with the world's second-largest nuclear arsenal) almost certainly responded in kind.

As a result, thousands of nuclear weapons stood poised on hair-trigger alert, ready to be launched by the president of either country with a decision time of just three minutes. The intercontinental nuclear-armed ballistic missiles controlled by these codes have a thirty-minute transit time from Russia to America and vice versa. Because they cannot be recalled, they pose an ever-present threat of global nuclear holocaust.

Aggressive militarization under the rubric of "defense against terrorism" threatens to provoke a chain reaction among nuclear nations, big and small, that, once set in motion, may prove impossible to control. No military confrontation anywhere in the world is free from this ominous and ever-present danger.

Weather as a Weapon: "Owning the Weather in 2025"

[Excerpts from a June 17, 1996, USAF Research Paper presented by Col. Tamzy J. House, Lt. Col. James B. Near Jr., LTC William B. Shields, Maj. Ronald J. Celentano, Maj. David M. Husband, Maj. Ann E. Mercer, and Maj. James E. Pugh.]

In 2025, U.S. aerospace forces can "own the weather" by capitalizing on emerging technologies and focusing development of those technologies to war-fighting applications. Such a capability offers the war fighter tools to shape the battlespace in ways never before possible. . . . A high-risk, high-reward endeavor, weather modification offers a dilemma not unlike the splitting of the atom.

While some segments of society will always be reluctant to examine controversial issues such as weather modification, the tremendous military capabilities that could result from this field are ignored at our own peril. . . . Weather-modification offers the war fighter a wide range of possible options to defeat or coerce an adversary.

In this paper we show that appropriate application of weather modification can provide battlespace dominance to a degree never before imagined. In the future, such operations will enhance air and space superiority and provide new options for battlespace shaping and battlespace awareness. The technology is there, waiting for us to pull it all together; in 2025, we can "Own the Weather."

Why would we want to mess with the weather? According to Gen. Gordon Sullivan, former Army chief of staff, "as we leap technology into the twenty-first century, we will be able to see the enemy day or night, in any weather—and go after him relentlessly."

A global, precise, real-time, robust, systematic weather-modification capability would provide war-fighting CINCs [commanders-in-chief] with a powerful force multiplier to achieve military objectives. Since weather will be common to all possible futures, a weather-modification capability would be universally applicable and have utility across the entire spectrum of conflict. The capability of influencing the weather even on a small scale could change it from a force degrader to a force multiplier.

People have always wanted to be able to do something about the weather. In 1957, the president's advisory committee on weather control explicitly recognized the military potential of weather modification, warning in their report that it could become a more important weapon than the atom bomb.

However, controversy since 1947 concerning the possible legal consequences arising from the deliberate alteration of large storm systems meant that little future experimentation could be conducted on storms, which had the potential to reach land. In 1977, the UN General Assembly adopted a resolution prohibiting the hostile use of environmental modification techniques. . . .

The influence of the weather on military operations has long been recognized. A significant number of [NATO's 1995] air sorties into Tuzla during the initial deployment supporting the Bosnian peace operation aborted due to weather.

Over 50 percent of the F-117 sorties weather aborted over their targets, and A-10s only flew 75 of 200 scheduled close air support (CAS) missions due to low cloud cover during the first two days of the campaign. The application of weather-modification technology to clear a hole over the targets long enough for F-117s to attack and place bombs on target or clear the fog from the runway at Tuzla would have been a very effective force multiplier.

Within the next three decades, the concept of weather modification could expand to include the ability to shape weather patterns by influencing their determining factors. By 2015, advances in computational capability, modeling techniques, and atmospheric information tracking will produce a highly accurate and reliable weather prediction capability, validated against real-world weather.

In the following decade, population densities will put pressure on the worldwide availability and cost of food and usable water. Massive life and property losses associated with natural weather disasters will become increasingly unacceptable. These highly accurate and reasonably precise civil applications of weather-modification technology have obvious military implications.

In the broadest sense, weather modification can be divided into two major categories: suppression and intensification of weather patterns. In extreme cases, it might involve the creation of completely new weather patterns, attenuation or control of severe storms, or even alteration of global climate on a far-reaching and/or long-lasting scale.

Extreme and controversial examples of weather modification—creation of made-to-order weather, large-scale climate modification, creation and/or control (or "steering") of severe storms, etc.—were researched as part of this study but receive only brief mention here because, in the authors' judgment, the technical obstacles preventing their application appear insurmountable within thirty years.

The number of specific intervention methodologies is limited only by the imagination, but with few exceptions they involve infusing either energy or chemicals into the meteorological process in the right way, at the right place and time. The intervention could be designed to modify the weather in a number of ways, such as influencing clouds and precipitation, storm intensity, climate, space, or fog.

Influencing precipitation could prove useful in two ways. First, enhancing precipitation could decrease the enemy's trafficability by muddying terrain, while also affecting their morale. Second, suppressing precipitation could increase friendly trafficability by drying out an otherwise muddied area. Research has been conducted in precipitation modification for many years, and an aspect of the resulting technology was applied to operations during the Vietnam War.

A pilot program known as Project Popeye conducted in 1966 attempted to extend the monsoon season in order to increase the amount of mud on the Ho Chi Minh trail, thereby reducing enemy movements. A silver iodide nuclei agent was dispersed from WC-130, F4, and A-1E aircraft into the clouds over portions of the trail winding from North Vietnam through Laos and Cambodia into South Vietnam. Positive results during this initial program led to continued operations from 1967 to 1972. . . .

International agreements have prevented the United States from investigating weather-modification operations that could have widespread, long-lasting, or severe effects. However, possibilities do exist (within the boundaries of established treaties) for using localized precipitation modification over the short term, with limited and potentially positive results.

Just as a black tar roof easily absorbs solar energy and subsequently radiates heat during a sunny day, carbon black also readily absorbs solar energy. When dispersed in microscopic or "dust" form in the air over a large body of water, the carbon becomes hot and heats the surrounding air, thereby increasing the amount of evaporation from the body of water below. . . .

Can this type of precipitation enhancement technology have military applications? Yes, if the right conditions exist. . . . Transporting it in a completely controlled, safe, cost-effective, and reliable manner requires innovation. Numerous dispersal techniques have already been studied, but the most convenient, safe, and cost-effective method discussed is the use of afterburner-type jet engines to generate carbon particles while flying through the targeted air. This method is based on injection of liquid hydrocarbon fuel into the afterburner's combustion gases.

Lab experiments have shown microwaves to be effective for the heat dissipation of fog. However, results also indicate that the energy levels required exceed the U.S. large power density exposure limit of 100 watt/m^2 and would be very expensive.

Field experiments with lasers have demonstrated the capability to dissipate warm fog at an airfield with zero visibility. Generating 1 watt/cm^2, which is approximately the U.S. large power density exposure limit, the system raised visibility to one-quarter of a mile in 20 seconds. . . . Recent Army research lab experiments have demonstrated the feasibility of generating fog. . . . This technology would enable a small military unit to avoid detection in the IR spectrum. Fog could be generated to quickly conceal the movement of tanks or infantry, or it could conceal military operations, facilities, or equipment.

The desirability to modify storms to support military objectives is the most aggressive and controversial type of weather modification. The damage caused by storms is indeed horrendous. For instance, a tropical storm has an energy equal to 10,000 one-megaton hydrogen bombs, and in 1992 Hurricane Andrew totally destroyed Homestead AFB, Florida, caused the evacuation of most military aircraft in the southeastern United States, and resulted in $15.5 billion of damage.

At any instant there are approximately 2,000 thunderstorms taking place. In fact 45,000 thunderstorms, which contain heavy rain, hail, microbursts, wind shear, and lightning form daily. . . . The danger of thunderstorms was clearly shown in August 1985 when a jumbo jet crashed, killing 137 people after encountering microburst wind shears during a rainsquall.

As indicated, the technical hurdles for storm development in support of military operations are obviously greater than enhancing precipitation or dispersing fog as described earlier. One area of storm research that would significantly benefit military operations is lightning modification. . . . Possible mechanisms to investigate would be ways to modify the electropotential characteristics over certain targets to induce lightning strikes on the desired targets as the storm passes over their location.

In summary, the ability to modify battlespace weather through storm cell triggering or enhancement would allow us to exploit the technological "weather" advances of our 2025 aircraft. . . .

Exploding/Disabling Space Assets Traversing Near-Space

The ionosphere could potentially be artificially charged or injected with radiation at a certain point so that it becomes inhospitable to satellites or other space structures. The result could range from temporarily disabling the target to its complete destruction via an induced explosion. . . .

Nanotechnology also offers possibilities for creating simulated weather. A cloud, or several clouds, of microscopic computer particles, all communicating with each other and with a larger control system, could provide tremendous capability. Interconnected, atmospherically buoyant, and having navigation capability in three dimensions, such clouds could be designed to have a wide range of properties. They might exclusively block optical sensors or could adjust to become impermeable to other surveillance methods. They could also provide an atmospheric electrical potential difference, which otherwise might not exist, to achieve precisely aimed and timed lightning strikes.

One major advantage of using simulated weather to achieve a desired effect is that unlike other approaches, it makes what are otherwise the results of deliberate actions appear to be the consequences of natural weather phenomena.

According to J. Storrs Hall, a scientist at Rutgers University conducting research on nanotechnology, production costs of these nanoparticles could be about the same price per pound as potatoes. This, of course, discounts research and development costs. . . .

Storm modification will become more valuable over time. The importance of precipitation modification is also likely to increase as usable water sources become more scarce in volatile parts of the world. The importance of space weather modification will grow with time.

The world's finite resources and continued needs will drive the desire to protect people and property and more efficiently use our croplands, forests, and range lands. The ability to modify the weather may be desirable both for economic and defense reasons. The global weather system has been described as a series of spheres or bubbles. Pushing down on one causes another to pop up. We need to know when another power "pushes" on a sphere in their region, and how that will affect either our own territory or areas of economic and political interest to the United States.

The lessons of history indicate a real weather-modification capability will eventually exist despite the risk. The drive exists. People have always wanted to control the weather, and their desire will compel them to collectively and continuously pursue their goal. The motivation exists. The potential benefits and power are extremely lucrative and alluring for those who have the resources to develop it. This combination of drive, motivation, and resources will eventually produce the technology.

History also teaches that we cannot afford to be without a weather-modification capability once the technology is developed and used by others. Even if we have no intention of using it, others will. To call upon the atomic weapon analogy again, we need to be able to deter or counter their capability with our own.

[*Read the full report at:* https://archive.org/stream/WeatherAsAForceMultiplier/WeatherAsAForceMultiplier_djvu.txt.]

Nuclear Doomsday

Daniel Ellsberg (2009)

One day in the spring of 1961, soon after my thirtieth birthday, I was shown how our world would end. Not the Earth, not (so far as I knew then) all humanity or life, but the destruction of most cities and people in the Northern Hemisphere. What I was handed, in a White House office, was a single sheet of paper with some numbers and lines on it. It was headed "Top Secret-Sensitive"; under that, "For the President's Eyes Only."

As a consultant to the Office of the Secretary of Defense, it was routine for me to read "Top Secret" documents. But I had never before seen one marked "For the President's Eyes Only," and I never did again.

A cover sheet identified it as the answer to a question President John F. Kennedy had addressed to the Joint Chiefs of Staff a week earlier. Bob Komer, the president's deputy assistant secretary of defense, showed it to me because I had drafted the question, which Komer had sent in the president's name.

The question to the JCS was: "If your plans for general [nuclear] war are carried out as planned, how many people will be killed in the Soviet Union and China?"

Their answer was in the form of a graph. The vertical axis was the number of deaths, in millions. The horizontal axis was time, indicated in months. The graph was a straight line, starting at time zero on the horizontal—on the vertical axis, the number of immediate deaths expected within hours of our attack—and slanting upward to a maximum at six months, an arbitrary cutoff for the deaths that would accumulate over time from initial injuries and from fallout radiation.

The lowest number, at the left of the graph, was 275 million deaths. The number at the right-hand side, at six months, was 325 million.

That same morning, with Komer's approval, I drafted another question to be sent to the Joint Chiefs over the president's signature, asking for a total breakdown of global deaths from our own attacks, to include not only the whole Sino-Soviet bloc but all other countries that would be affected

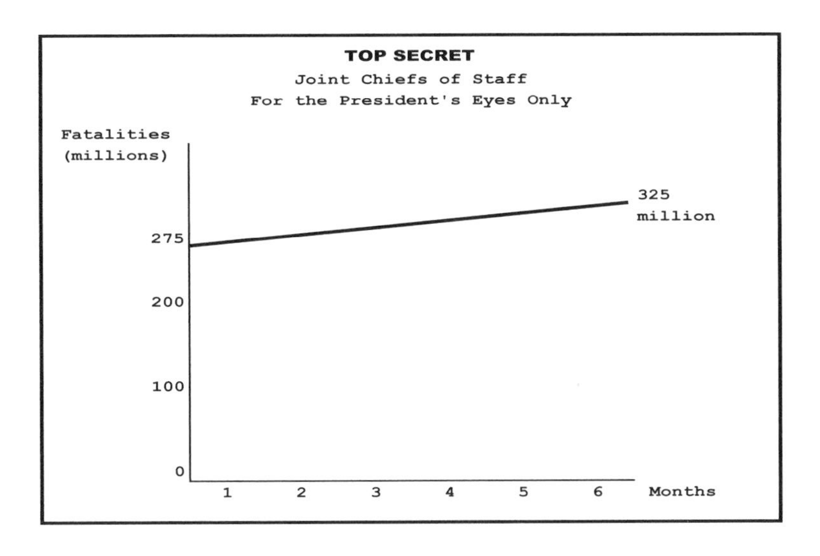
TOP SECRET
Joint Chiefs of Staff
For the President's Eyes Only

by fallout. Again their answer was prompt. Komer showed it to me about a week later, this time in the form of a table with explanatory footnotes.

In sum, 100 million more deaths, roughly, were predicted in East Europe. There might be an additional 100 million from fallout in West Europe, depending on which way the wind blew (a matter, largely, of the season). (The total number of "casualties"—injured as well as killed—had not been requested and was not estimated. Nor were casualties from any Soviet retaliatory strikes.)

The total death toll as calculated by the Joint Chiefs, from a U.S. first strike aimed primarily at the Soviet Union and China, would be roughly 600 million dead. A hundred holocausts.

I remember what I thought when I held the single sheet with the graph on it. I thought, this piece of paper should not exist. It should never have existed. Not in America. Not anywhere, ever. It depicted evil beyond any human project that had ever existed. There should be nothing on Earth, nothing real, that it referred to.

But I knew what it dealt with was all too real. I had seen some of the smaller bombs myself, H-bombs with an explosive yield of 1.1 megatons each—equivalent to 1.1 million tons of high explosive, each bomb half the total explosive power of all the bombs of World War II combined. I saw them slung under single-pilot F-100 fighter-bombers on alert at Kadena

Air Base on Okinawa, ready to take off on ten minutes' notice. On one occasion, I had laid my hand on one of these, not yet loaded on a plane. On a cool day, the smooth metallic surface of the bomb was warm from the radiation within: a body-like warmth.

A Death Notice for a Billion People

In the bomb bays of the SAC [Strategic Air Command] planes were thermonuclear bombs much larger than those I saw in Okinawa. Many were from five to twenty megatons in yield. Each 20-megaton bomb—1,000 times the yield of the fission bomb that destroyed Nagasaki—was the equivalent of 20 million tons of TNT, or 10 times the total tonnage the United States dropped in World War II. Some 500 bombs in the arsenal each had the explosive power of 25 megatons. Each of these warheads had more power than all the bombs and shells exploded in all the wars of human history.

The planned targets for the whole force included, along with military sites, every city in the Soviet Union and China. One of the principal expected effects of this plan—partly intended, partly (in allied, neutral, and "satellite" countries) unavoidable "collateral damage"—was summarized on the piece of paper I held that day in the spring of 1961: the extermination of over half a billion people.

In fact, this was certainly a vast underestimate of the fatalities. Dr. Lynn Eden, a scholar at Stanford's Center for International Security and Cooperation, has revealed in *Whole World on Fire* (Cornell, 2004) the bizarre fact that the war planners of SAC and the Joint Chiefs have—throughout the nuclear era, to the present day—deliberately omitted entirely from their estimates of the destructive effects of U.S. or Russian nuclear attacks the effects of *fire*. They have done so on the grounds that these effects are harder to predict than the effects of blast or fallout on which their estimates of fatalities are exclusively based. Yet the firestorms caused by thermonuclear weapons are known to be predictably the largest producers of fatalities in a nuclear war! Given that for almost all strategic nuclear weapons the damage radius of firestorms would be two to five times the radius destroyed by blast, a more realistic estimate of the fatalities caused directly by the planned U.S. attacks would surely have been double the figure on the summary I held in my hand—a billion people or more.

Deterring Soviet non-nuclear aggression in Europe—say, a military occupation of West Berlin—depended ultimately on a presidential commitment to direct, if necessary, a U.S. nuclear first strike on the Soviet Union. All large cities of both the Soviet Union and China (even if China had no part in the crisis or hostilities triggering execution of this plan) were high on the list for initial, simultaneous missile attacks, and for subsequent coverage by bombers—along with the highest-priority Soviet missile sites, air bases, air defenses, and command centers.

The Thermonuclear Mousetrap

In the White House in January 1961, I had informed the newly arrived assistant to the president for national security, McGeorge Bundy, of a number of little-known facts and problems. One of these was the focus on U.S. first-strike plans in American preparations for any conflict with the Soviet Union involving forces above the level of a brigade. Another was Eisenhower's approval of operational planning to destroy an "optimum mix" of population targets along with military sites no matter how the conflict had originated.

A third subject in my briefing was the variety of ways in which the strategic forces might be triggered "by accident": by false alarm, miscalculation, miscommunication, or actions not directly authorized by the president or perhaps by any high-level commander.

The last point in particular caught Bundy's attention. I reported what I had learned in the Pacific, one of the most sensitive secrets in the system: that to forestall the possibility that our retaliatory response might be paralyzed either by a Soviet attack on Washington or by presidential incapacity, President Eisenhower had, as of 1958, secretly delegated to theater commanders the authority to launch nuclear operations in a crisis, either in the event of the physical unavailability of the president—Eisenhower himself had suffered both a stroke and a heart attack in office—or if communications with Washington were cut off.

The combined message of these reports was that our overall system for strategic response had the character of a giant thermonuclear mousetrap on a hair trigger. For a wide variety of provocative circumstances—definitely not requiring and most not involving either Soviet-initiated nuclear attacks or imminent expectation of them—it was set inflexibly to annihilate a large fraction of the civilian population of the Soviet Union and China, and of many allies and neutrals.

I myself at that time was neither a pacifist nor a critic of the explicit logic of deterrence or its legitimacy. On the contrary, I had been urgently working with my colleagues to assure a survivable U.S. capability to threaten clearly unacceptable damage to the Soviet Union in response to the most successful possible Soviet nuclear attack on the United States. But planned slaughter of 600 million civilians—10 times the total death count in World War II, a hundred times the scale of the Holocaust? That aimed-for accomplishment exposed a dizzying irrationality, madness, insanity, at the heart and soul of our nuclear planning and apparatus.

The chart I held in my hand that spring morning said to me that any confidence—worse, it seemed, any realistic hope—that the alert forces on either side might never be used was ill-founded.

The Americans who had built this machine, knowing, it turned out, that it would kill more than half a billion people if it were turned on—and who were unabashed in reporting this to the president—humans like that would not fail to pull the switch if ordered to do so by a president, or possibly by a superior other than the president.

And the presidents themselves? A few months earlier, Dwight Eisenhower had secretly endorsed the blueprints of this multi-genocide machine. He had furthermore demanded (largely for budgetary reasons) that there be *no other plan* for fighting Russians. He had approved this single strategic operational plan despite reportedly being privately appalled by its implications. And the Joint Chiefs had responded so promptly to his successor's question about the human impact of our planned attacks because they clearly assumed that John Kennedy would not, in response, order them to resign or be dishonorably discharged, or order the machine to be dismantled. (In that, it turned out, they were right.)

Surely neither of these presidents actually desired ever to order the execution of these plans, nor would any likely successor want to take such an action. But they must have been, or should have been, aware of the dangers of allowing such a system to exist. They should have reflected on, and trembled before, the array of contingencies—accidents, false alarms, outages of communications, Soviet actions misinterpreted by lower commanders, unauthorized action—that might release pent-up forces beyond their control; and on possible developments that could lead them personally to escalate or launch a preemptive attack.

Eisenhower had chosen to accept these risks. To impose them on humanity, and all other forms of life. Kennedy and Lyndon B. Johnson, to my direct knowledge, did likewise. So did Richard Nixon. To bring this story up to the present, there is much evidence—and none to the contrary—that the same has been true of every subsequent president.

Nuclear Winter

Moreover, the scale of the potential catastrophe was and remains vastly greater than I or the JCS or any presidents imagined over the next twenty years. Not until 1982–83 did new calculations—reconfirmed by Rutgers University in 2014—reveal that hemispheric and possibly global clouds of smoke and soot from the burning cities attacked by U.S. or Russian forces would block out sunlight for a prolonged period, lowering temperature drastically during spring and summer, freezing lakes and rivers and destroying crops worldwide. This "nuclear winter" could extinguish many forms of life and lead to the starvation deaths of billions of humans.

Yet the "option" of massive attacks on cities (or, euphemistically, upon industrial and military targets within or near cities) almost surely remains one among many planned alternatives, ready as ever to be carried out, within the strategic repertoire of U.S. and Russian plans and force readiness: this, a quarter-century after the discovery of the nuclear winter phenomenon.

A 2007 peer-reviewed study in *Atmospheric Chemistry and Physics* (a journal published by the European Geosciences Union) concluded that "the estimated quantities of smoke generated by attacks totaling little more than one megaton of nuclear explosives [two countries launching fifty Hiroshima-size bombs each] could lead to global climate anomalies exceeding any changes experienced in recorded history. The current global arsenal is about 5,000 megatons." The 2002 Strategic Offensive Reductions Treaty called for the United States and Russia each to limit their operationally deployed warheads to between 1,700 and 2,200 by December 2012, but a December 2008 study in *Physics Today* estimates that "the direct effects of using the 2012 arsenals would lead to hundreds of millions of fatalities. The indirect effects [long-term, from smoke] would likely eliminate the majority of the human population."

It is the long-neglected duty of the American Congress to test these scientific findings against the realities of our secret war plans. It is Congress's responsibility to investigate the nature of the planned targets for the reduced operational forces proposed by Obama and former Russian prime minister Dmitry Medvedev—1,500 to 1,675—or some lower but still huge number like 1,000, and the foreseeable human and environmental consequences of destroying those targets with the attacks currently programmed.

I knew personally many of the American planners, though apparently—from the fatality chart—not quite as well as I had thought. What was frightening was precisely that I knew they were not evil, in any ordinary, or extraordinary, sense. They were ordinary Americans, capable, conscientious, and patriotic. I was sure they were not different, surely not worse, than the people in Russia who were doing the same work, or the people who would sit at the same desks in later U.S. administrations.

That chart set me the problem, which I have worked on for nearly half a century, of understanding my fellow humans—us, I don't separate myself—in the light of this real potential for self-destruction of our species and of most others. Looking at the steady failure in the two decades since the ending of the Cold War to reverse course or to eliminate this potential, it is hard for me to avoid concluding that this potential is more likely than not to be realized in the long run.

The more one learns about the hidden history of the nuclear era, the more miraculous it seems that the doomsday machines that we and the Russians have built and maintained have not yet triggered each other. At the same time, the clearer it becomes that we could and that we must dismantle them.

THE AFTERMATH

I dream of giving birth to a child who will ask, "Mother, what was war?"

—Eve Merriam, U.S. poet and playwright

Graveyards, Waste, and War Junk

Susan D. Lanier-Graham (2017)

While environmental destruction during warfare is often overlooked as the more immediate concerns of human death and the loss of homes take center stage, after a war is over and countries are left to clean up the damage, ecological devastation often becomes more noticeable.

The abandonment of materials and explosives can be devastating. This might include battle areas strewn with supplies, debris, trash, and perhaps even toxic chemicals. A major threat following warfare is the explosion of undetected munitions or the presence of radioactive residues on the battlefield. Further destruction can result from deforestation and erosion that worsens over time, long after the fighting ceases. Often the gradual extinction of wildlife species is the last and most complete loss to an already devastated ecosystem.

Hollywood has made the image of a battlefield a familiar sight. Visions of cannon, guns, canteens, bombed-out tanks, broken planes, and sunken ships are images that the big screen frequently has utilized. Hollywood's war films never romanticize or dwell upon the hastily constructed buildings, the bunkers, the trash left after the fighting has stopped. Yet these "leftovers" of war are often among the most troublesome problems facing a recovering nation.

Let's look first at examples from early American history.

After large battlefield clashes, bodies of the dead (both men and horses) were frequently left lying on the bloodied ground. (This was especially a problem in the summer, when the heat only makes the situation more serious.) In both in the American Revolution and the Civil War, bodies were actually dumped in creeks to get rid of them, a practice that tainted the water supply for everyone downstream.

Combat also left debris scattered across the battlefields. Civil War-era equipment was heavy and cumbersome. If a retreating army had to leave an area quickly (or if too many troops were killed), much of the larger pieces of equipment had to be abandoned. One way to ensure that the enemy could never again utilize a canon abandoned on the battlefield was to drive a spike through the canon's iron muzzle, rendering the weapon unusable. The sabotaged weapon simply remained in place at the point where it had last been fired.

Since the Revolution, the remains of sunken ships have been resting at the bottom of America's rivers, seas, and oceans. The bottom of the York River, at the mouth of the Chesapeake Bay, is the grave of more than fifty warships sunk during the siege of Yorktown at the end of the Revolutionary War. During the Civil War, ships were sunk off the Atlantic Coast and along the expanse of the Mississippi River. Perhaps the most famous relic of the Civil War is the *USS Monitor,* a 172-foot-long ironclad that was sunk off the North Carolina coast.

During World War I, boats littered the floor of the Atlantic Ocean, along with their cargo. More devastating than the ships, however, were the substances purposely dumped into the ocean following World War I. At the end of the war, the Germans were ordered to dump 20,000 tons of stockpiled chemical weapons into the Baltic Sea. Since that time, Danish fishermen have occasionally been poisoned by mustard gas leaking from rusty containers and into the sea. *Der Spiegel* reported in 2013 that "more than 50 million bombs, shells, detonators, and cartridges from World War II are rusting away on the floor of the North and Baltic Seas or are washing up on beaches."

In addition to the sunken ships that went to the bottom during the naval battles of World War II, hundreds of aircraft—from bombers to fighters shot down during combat missions—now rest beneath the oceans of the South Pacific.

The Poisoning of the Skagerrak Seabed

Aside from the aesthetic offense of half-submerged ships littering once-pristine waters, these visible reminders of battles past continue to pose a serious threat to marine life owing to the hazardous nature of the substances found onboard. Chemical leaks and oil spills from sunken ships can cause severe damage to marine life and life along nearby shores.

A vivid example is found in the Skagerrak seabed between Norway, Sweden, and Denmark. It is known that at least twenty-one German ships (some estimates run as high as fifty vessels) loaded with chemical weapons were sunk in the area. An allied commission formed at the end of World War II decided that the quickest and cheapest way to dispose of the 300,000-plus tons of captured German chemical weapons was to load them on damaged ships and sink them.

A 1984 report by the Danish Ministry of the Environment claimed that between 36,000 and 50,000 tons of chemical ammunition had been dumped in the Baltic Sea by Soviet Occupation forces following World War II. It was also discovered that the Germans sunk a ship containing 5,000 tons of chemical weapons off the coast of Denmark just before the war ended. The Danish report further documented that 34 ships and 151,425 tons of ammunition had been sunk in the Skagerrak.

Some leaking containers have been retrieved. Unfortunately, much of the recovered mustard gas is still potent and remains intact inside the disintegrating containers. The chemical reactions that occur as the mustard gas breaks down can produce deadly byproducts that are more toxic than the original gas. Experts suggest these reactions may continue for as long as 400 years.

Unexploded Ordnance

One of the most devastating and long-term consequences of modern war is unexploded ordinance. While unexploded munitions pose a grave threat to humans, they are also a danger to the natural environment and to native wildlife.

Hidden and undetected, unexploded artillery and mortar shells, anti-tank and antipersonnel mines, bombs, and grenades can remain active for many years. Even today, there are shells and grenades remaining on the World War I site of the Battle of Verdun, which took place more than a century ago.

During World War II, it has been calculated, the Germans and the Allies mined 80 percent of the entire landmass of Poland. According to a 1985 report presented to the United Nations Environmental Programme (UNEP), 15 million mines had been discovered in Poland—and more continue to be discovered on a regular basis.

Since the end of World War II, 16 million buried artillery shells, 490,000 bombs, and 600,000 underwater mines have been discovered in France. Because of the staggering number of discoveries, many stretches of the French countryside remain fenced off due to lingering risks.

In 2013, 20,000 residents of Dortmund, Germany, were forced from their homes after the discovery of a buried British Royal Air Force "blockbuster" bomb. In 2014, another World War II–era bomb killed an excavator operator in North Rhine-Westphalia. On Christmas Day 2016, 54,000 terrified people fled their apartments after a 1.8-ton British "mega bomb" was discovered in the center of Augsburg, Austria's third-largest city.

In August 2015, a 550-pound German bomb was found in the basement of a three-story building in East London. Two months later, another unexploded bomb forced the evacuation of a market in nearby Spitalfields.

In November 2016, a World War II–era bomb was unearthed during construction near Florida State University at the site of a long-abandoned military airfield. The bomb was safely detonated with a blast that shook local shops and the state capitol.

And in February 2017, 72,000 Greek citizens were hastily evacuated after a 500-pound U.S. gravity bomb (detonator still intact) was discovered buried beneath a gas station.

Another heavily mined area was Libya, where an estimated five million mines were placed. The areas of Libya most heavily mined were large tracts of arid land and rangeland. An estimated 125,000 domestic animals and an undetermined number of wild animals were killed as a result of encounters with unexploded ordinance. Sixty percent of the victims were camels.

In addition to the tragic loss of life that accompanies these unexpected explosions, damage to the soil can pose a long-term danger from erosion. When vegetation is destroyed, wildlife habitat is threatened. According to UNEP, gazelles in North Africa no longer live in areas that were heavily mined during World War II.

Another danger from unexploded munitions lies at the bottom of various seas, ports, and rivers throughout the world. World War II–era sea mines have been found at the approaches to the port of Le Havre, France, in the waterways of Berlin, in the Thames estuary in the United Kingdom, and even some sixty miles off the coast of North Carolina. The biggest danger to the environment (even greater than the danger from an actual underwater explosion) is the explosive material found in the munitions. Cyclonite, a common military explosive, is also a potent nerve poison, toxic to mammals and used commercially as a rat killer. As a poison, cyclonite remains active in seawater for many years without dissipating.

Deadly Debris Lingers for Decades

Modern technology and greater numbers of personnel and weapons have made the problems of warfare more intense and longer lasting. In 1975, military historian William Bartsch made a trip through the Gilbert and Ellice Islands doing research on the remains from the war in the Pacific.

On the island of Funafuti, Bartsch found the "borrow pits" to be the most common reminders of the war. These large pits were dug to excavate coral for use in constructing runways. They eventually became dumps for withdrawing U.S. troops at the end of the war. Today, these pits stand at each end of the island and provide an open grave for boiler tanks, vehicle chassis, tires, spare parts, and other junk the United States no longer had any use for.

On Betio Island, war remains are visible everywhere, resting on the once-pristine beaches. Artifacts include an entire Sherman tank, bunkers, concrete blocks used to divert landing craft at the beaches' edge, and, occasionally, live munitions.

Poignant reminders of the war are also found in the Aleutian Islands and lower Alaska Peninsula. According to an Army Corps of Engineers' 1979 Draft Environmental Impact Statement, the debris consists of

> the remains of troop quarters, mess halls, gymnasiums, warehouses, power plants with engines and generators, ammunition magazines and bomb dumps, fuel depots, garages and workshops, airplane runways and hangars, hospitals, radio and weather stations, gun emplacements, bunkers, and miscellaneous material including live and detonated ordnance, vehicles and heavy machinery, pierced steel air strip matting, barbed wire, communications and utility poles and cable, pipelines, anti-submarine nets, bedsprings, and fifty-five-gallon POL [petroleum, oils, lubricants] drums.

War debris on the Aleutian Islands has contaminated the islands' water supplies. Asphalt, grease, motor oil, paint, paper, fiberboard, plasterboard, rusting drums, sheet metal, tires, cable, insulating materials, and explosives and gunpowder have been found in the waters off the Aleutian Islands.

The people of Vietnam have faced multiple postwar dilemmas. Numerous tanks, airplanes, and other remnants of the war are still evident today, all of which will remain unless physically removed. Sites where U.S. bases were located contain the remains of a modern civilization at war—petroleum products, solvents, toxins used in the defoliation program, and numerous other poisons, while antitank and antipersonnel mines remain hidden deep in the jungles of Southeast Asia.

Thousands of tons of unexploded ordinance were left behind after the first Persian Gulf War. In the deserts of Kuwait and Iraq mines were planted in soft sand that blows and shifts continuously. Blowing and drifting sands make it impossible to flag unexploded bombs for removal. The sands may cover the dangerous object completely, only to uncover it again years later.

A U.S. government study estimated that 70 percent of the conventional bombs dropped over Iraq missed their target. Of the 88,500 tons of bombs dropped on Iraq, 17,700 tons, or 20 percent of the total, never exploded.

The Persian Gulf War was somewhat different where waste disposal was concerned. The amount of trash generated by a force of close to half a million hungry soldiers is astronomical. It has been estimated that U.S. troops—eating two meals each day from the plastic MRE (Meals Ready to Eat) bags—discarded 6 million used plastic bags each week.

Recycling efforts are not at the forefront of concerns during a war. Water comes in plastic bottles, soft drink cans are aluminum, junk food is wrapped in cellophane and cardboard. In Kuwait and Saudi Arabia, a good portion of this postwar trash remained in the desert—the responsibility of Saudi Arabia and Kuwait to clean up. Also left behind: the usual military wastes—solvents, acids, paint, fuel, lubricants, explosives—all capable of polluting underground aquifers critical to the desert ecosystem.

The standard post-conflict solution has been to bury what could be buried, filling huge, hastily dug landfills. The burial sites often were left exposed, with huge dumps of trash and debris marking the American exit from the combat zone. What could not be buried was usually incinerated, often producing a peculiar stench that soldiers seem to remember forever.

The Pentagon's Toxic Burn Pits

Katie Drummond (2013)

At the height of the war in Iraq, U.S. forces operated out of 505 bases scattered across the country. Joint Base Balad, a fifteen-square-mile outpost north of Baghdad, was the second largest. Home to 36,000 military personnel and contractors at its peak, the base was considered a vital hub for operations throughout Iraq—largely thanks to two 11,000-foot runways and one of the best and biggest trauma centers in the region. Balad also boasted a notorious array of amenities: troops living in the makeshift mini-city could dine at Burger King or Subway, play miniature golf or relax in an air-conditioned movie theater, and browse for TVs or iPods at two different shopping centers.

But when Le Roy Torres arrived at Balad in the summer of 2007, the first thing he noticed was the smell—a noxious, overwhelming stench reminiscent of burning rubber. "I was like, 'Wow, that is something really bad, really, really bad,'" he recalls. Soon, he also noticed the smoke: plumes of it curling into the air at all hours of the day, sometimes lingering over the base as dark, foreboding clouds. That smoke, Le Roy soon learned, was coming from the same place as the stench that had first grabbed him: Balad's open-air burn pit.

The pit, a shallow excavation measuring a gargantuan ten acres, was used to incinerate every single piece of refuse generated by Balad's thousands of residents. That meant seemingly innocuous items, like food scraps or paper. But it also meant plastic, Styrofoam, electronics, metal cans, rubber tires, ammunition, explosives, human feces, animal carcasses, lithium batteries, asbestos insulation, and human body parts—all of it doused in jet fuel and lit on fire.

The pit wasn't unique to Balad: open-air burn pits, operated either by service members or contractors, were used to dispose of trash at bases all across Iraq and Afghanistan.

"I remember waking up with soot on me; you'd come out and barely see the sun because it was so dark from the smoke," says Dan Meyer, a twenty-eight-year-old Air Force veteran who lived adjacent to the burn pit at Afghanistan's Kandahar Air Base. Meyer is now confined to a wheelchair because of inoperable tumors in his knees, and breathes using an oxygen tank due to an obstructive lung disease. "It would just rain down on us. We always called it 'black snow.'"

It's no secret that open-air burning poses health hazards. The Environmental Protection Agency has long warned that burning waste—even organic refuse like brush or tree branches—is dangerous. Burning items like plastic water bottles or computer parts is even worse. "It's appalling," says Anthony Wexler, PhD, director of the Air Quality Research Center at UC Davis and the coauthor of a 2010 review of the military's air-quality surveillance programs in Iraq and Afghanistan. "From a health perspective, this kind of open-pit burning, especially when you're burning everything under the sun, creates a real mess." That's because of both the size of the particulate matter emitted from the pits and its composition. Smoke from any combustion process fills the air with what are known as "fine particles" or PM2.5. Because they're so small—measuring 2.5 microns in diameter or less—these particles burrow more deeply into the lungs than larger airborne pollutants, and from there can leach into the bloodstream and circulate through the body. The military's burn pits emitted particulate matter laced with heavy metals and toxins—like sulfur dioxide, arsenic, dioxins, and hydrochloric acid—that are linked to serious health ailments. Among them are chronic respiratory and cardiovascular problems, allergies, neurological conditions, several kinds of cancer, and weakened immune systems.

Le Roy is convinced that burn-pit exposure is behind his health problems, which he says first emerged a few weeks into that 2007 deployment. "It started with a cough. I was coughing up this gunk stuff, like black phlegm that kept coming and coming," he says. "The medical officer told me it was 'Iraqi crud' and it'd go away in a few days. I thought, 'I've been here a month, how much longer?'" The cough never improved, and upon his return to the United States in 2008, Le Roy found himself struggling to get answers from military physicians. They brushed it off as bronchitis, asthma, even an anxiety disorder triggering physical symptoms.

As a reservist, Le Roy also had a civilian job as a Texas state trooper. But despite fourteen years in the role, he was put on leave from his job in September 2010, two years after returning from Iraq, for being unable to perform physically challenging tasks.

Later that same year, Le Roy was referred to Dr. Robert Miller, a pulmonologist at Vanderbilt University. Dr. Miller had met soldiers like Le Roy before—dozens of men and women with chronic respiratory problems following deployments to Iraq or Afghanistan—and he knew exactly what to do: Le Roy soon underwent a lung biopsy, a procedure wherein surgeons make three incisions in the chest to remove tissue for examination. For Le Roy, much like Dr. Miller's other patients, the diagnosis was grim: he suffered from constrictive bronchiolitis, an exceedingly rare, sometimes terminal lung disease.

Turning a Blind Eye to Burn-Pit Toxins

The military has long known, at least internally, that burn pits can harm human health. In a series of waste-management guidelines published in 1978, the DOD cautioned that open-air burning was not a safe option for waste disposal, and should only be used "[when] there is no other alternative." There's no doubt that burn pits are an expedient, inexpensive way to dispose of trash—especially in the early phases of a conflict—but replacements like closed incinerators or landfills offer a longer-term solution once a base is established. A few years into the wars in Iraq and Afghanistan, however, some military personnel warned that those replacements weren't being implemented.

"[Burn pits] should only be used in the interim until other ways of disposal can be found," notes a 2006 memo from the now-retired Lt. Colonel Darrin Curtis, then a bioenvironmental engineer with the Air Force. "It is amazing that the burn pit [at Balad] has been able to operate without restrictions over the past few years." Another memo, this one from 2011, warns of "an increased risk of long-term adverse health conditions" caused by the burn pit at Bagram Air Base in Afghanistan.

Meanwhile, some civilian experts like Dr. Miller were seeing patients with worrisome medical problems. "It began to look to me like a perfect storm," he recalls. "We had one soldier after another—talented, capable athletes—who couldn't pass their fitness tests anymore." But in 2010, after he warned the military that dozens of his patients had constrictive bronchiolitis—which is almost always caused by toxic exposure in otherwise healthy people—a troubling thing happened: physicians at Fort Campbell were directed to stop referring soldiers to Dr. Miller's nearby medical practice. "They basically cut us off," he says. "Needless to say, it's clear that the DOD hasn't embraced this as a significant problem."

Eventually, ongoing pleas from soldiers and doctors garnered attention from politicians and military leadership. In 2009, the U.S. government passed legislation to ban the open-air burning of some trash, namely "hazardous or biomedical waste," and U.S. Central Command ordered the closure of burn pits on bases with more than 100 soldiers, "when the transition is practical." The military gradually replaced some burn pits with closed incinerators. But by many accounts, this didn't remedy the problem comprehensively or quickly enough. A 2010 GAO report found that some pits in Iraq were still burning items, like plastics and Styrofoam, banned by those new federal rules. And as recently as July 2013, four pits in Afghanistan were still operating—even as closed incinerators built on two of those bases, at a cost of $16 million, collect dust.

"I took it personally, that they made a calculation about what was a cheaper and easier way to get rid of trash, versus the cost of someone's life,"

Le Roy says. "Here someone is worrying about mortars or IED attacks, and in the end, it was our own guys who got us."

America's "Downwinders"

James Lerager (2016)

For close to half a century, the United States government tested nuclear weapons at the Nevada Test Site—a 1,375-square-mile expanse of desert located approximately 60 miles north of Las Vegas. According to government records, exactly 100 atomic bombs were exploded in the air above the site between 1951 and 1962. After the signing of the Limited Nuclear Test Ban Treaty in 1963, the United States detonated an additional 800 nuclear weapons and devices underground—some of them 10 times more powerful than the bomb dropped on Hiroshima.

Until 1992, when Congress legislated a moratorium on nuclear weapons testing, the region was regularly shaken by earthquake-like shockwaves. Throughout the 1950s, residents of Nevada and Utah frequently observed predawn flashes of light across the distant horizon; indeed, during the years of aboveground testing, the government actively encouraged civilians to observe the nuclear explosions from designated vista points. And throughout these decades, the agencies that conducted the testing—the Atomic Energy Commission (later to become the Department of Energy) and the Department of Defense—assured the American people that atmospheric nuclear testing had been completely safe and that underground testing was even safer.

The government timed its atmospheric nuclear explosions carefully, scheduling them on days when officials expected prevailing winds to blow the fallout clouds to the east and southeast—toward St. George, Utah, or toward the Four Corners, a region heavily populated by Navajo, Hopi, and Pueblo Indians. As an intended result, Mormon and Hispanic communities, and Indian reservations, repeatedly received the bulk of the radioactive fallout. Las Vegas, home to most of the tens of thousands of scientists, engineers, and construction workers who worked at the test site, was intentionally protected from the fallout clouds, although the predawn and early morning nuclear detonations were often visible from the city.

Inevitably, however, things didn't always go according to plan. Sometimes the winds unexpectedly shifted direction during or immediately after an explosion, blowing the fallout clouds not east, but west toward California. On at least two documented occasions during the 1950s, radioactive clouds settled over the cities of Riverside and Los Angeles, releasing their dangerous burden on unsuspecting residents. Occasionally the fallout blew across the Midwest, even to New England. Whether in Utah, California, or New York, civilian populations rarely received warnings when fallout struck their communities.

Although the region's permanent population was relatively small in the '50s, the Southwest was America's summer vacation heartland, hosting countless station wagons packed with children visiting "Indian Country" and the Canyonlands. Families from across the country camped and hiked the red rock of Zion, Bryce, Grand Canyon, Monument Valley, and the other great parks of the Southwest. They visited the tourist meccas of Santa Fe, Salt Lake City, and Las Vegas. Millions of Americans crisscrossed the region even as radioactive clouds moved silently across the desert and through the canyons.

Dead Sheep and Lawsuits

By the early 1950s, AEC and Defense Department officials were fully aware that even low doses of radiation were potentially dangerous, but would not be immediately apparent; increases in birth anomalies would occur months later or years later, and cancers would take years, even decades, to develop. But the government seemingly was more interested in protecting its nuclear weapons testing program from scrutiny than in protecting these downwind people from radiation. On the rare occasions that the Atomic Energy Commission issued local fallout warnings, it also told residents that any radioactive clouds would pass quickly, causing no lasting harm. According to AEC advisories, people living downwind needed to remain indoors for only a few hours as the fallout clouds passed before resuming daily activities.

However, fallout quickly and dramatically affected the downwind ranchers in Nevada and Utah. Soon after wind-borne clouds of radiation swept over the open ranges, the ranchers' sheep, cattle, and horses developed sores on their mouths and faces. The animals' coats soon began to fall out in patches and some animals died. When spring arrived, pregnant sheep, cattle, and horses often bore malformed or sickly offspring. Sometimes hundreds of animals mysteriously sickened and died. The ranchers themselves occasionally became nauseous after the fallout clouds passed over, and sometimes they temporarily lost their own hair, a telling signal of acute radiation illness.

By the mid-1950s, ranchers came to believe that fallout from the Nevada Test Site was causing previously unseen and unfamiliar yet widespread illness and death in their herds. Some ranchers brought lawsuits against the government. In court, scientists and technicians from the AEC argued against the accusations, swearing that the tests were completely safe—and sometimes suggesting that the animals' plight demonstrated the ranchers' incompetence.

The AEC consistently prevailed in those initial court cases—because judges and jurors believed the AEC scientist's testimony was credible. In 1979, however, declassified AEC documents presented during a Congressional hearing revealed that high-level AEC officials had ordered agency scientists to present deceptive testimony to the courts—even when those employees knew that the nuclear tests and resultant fallout were inherently dangerous.

Uranium Mining

The Southwest was also home to an earlier phase of the nuclear weapons cycle: the mining of uranium. Hundreds of underground and open-pit mines scar the deserts and mountains of Nevada, Utah, Arizona, Colorado, and New Mexico. In the 1940s and '50s, these states experienced a "uranium rush" reminiscent of nineteenth-century gold and silver rushes. Although not currently operating—uranium demand and prices fell precipitously during the 1980s and '90s after the Three Mile Island and Chernobyl nuclear accidents—mining claims remain active and uranium mines could quickly reopen. In 2008, as world uranium prices began rising once again, both mining companies and wildcatters started staking out thousands of new uranium claims on public lands throughout the American West.

More than a half-century later, the toxic legacy from the 1950s boom persists—in the form of countless mounds of toxic uranium mine waste tailings scattered across the Southwest and at hundreds of sites on the Navajo reservation. These mining wastes continue to spread as wind-borne dust washes into rivers and streams or falls on topsoil where it leaches into the subterranean aquifers that are essential for the region's water supply. Uranium mining wastes and tailings ponds from the Atlas Uranium Mill, near Moab, Utah, are washing into the Colorado River (the tailings are just 750 feet from the river). A proper stabilization and cleanup program may ultimately cost hundreds of millions of taxpayer dollars.

The 3,000-acre Jackpile Mine (once the world's largest open-pit uranium operation) contained three open pits and 32 waste dumps. Although the mine closed in 1979, its wastes continued to poison the Laguna Indian reservation west of Albuquerque, New Mexico. Atlantic Richfield Corporation (ARCO) spent $48 million for a partial cleanup at the Jackpile Mine. The Jackpile Reclamation Project was officially completed in 1995, but an EPA reassessment in 2007 determined that radiation was still being released. The estimated cost for remediation to the fullest extent possible has been set at $400 million.

The uranium mines were often primitive, poorly ventilated, unsafe, and filled with invisible radioactive gases. Many of the people who worked the mines were Native Americans, often from the Navajo Nation. A significant but unknown percentage of these miners later developed lung and other cancers from their exposure to radiation in the mines and a significant but also unknown number of their children developed leukemia and other complex health problems—facts that the government belatedly acknowledged in the late 1980s.

In 1990, Congress passed the Radiation Exposure Compensation Act (RECA) and issued a formal apology to the people of the Southwest for the physical suffering inflicted on them from nuclear fallout and work in the uranium mines—a suffering they had been forced to endure for four decades without official acknowledgment or aid. Congress publicly admitted that federal agencies had intentionally deceived the public and the courts regarding potential dangers from exposure to fallout and from working in unsafe and poorly ventilated uranium mines. The RECA established a $200 million compensation fund for those uranium miners and downwinders who could demonstrate radiation injury—as defined within the restricted terms of the act. The RECA's list of covered injuries includes lung cancer and several other radiation-related classes of cancer. (By way of cost comparison, each of the hundreds of nuclear tests conducted at the Nevada Test Site cost in excess of $100 million.)

Radioactive contamination from the drifting fallout clouds of the 1950s has receded, but even today all is not quiet in the miles of underground tunnels and chambers on the Nevada Test Site. Since 2001, more than a dozen subcritical nuclear tests (using small quantities of plutonium detonated with chemical explosives, simulating the signature of a nuclear detonation) have been conducted at the site. In 1996, negotiators from around the world agreed on a long-awaited Comprehensive Nuclear Test Ban Treaty. Although President Clinton urged its approval, the U.S. Senate voted to reject the treaty in 1999. President George W. Bush and his administration actively opposed ratification of the treaty and further alarmed the world by announcing its decision to maintain the Nevada Test Site in a state of readiness to resume full-scale underground nuclear testing. President Barack Obama's 2010 Nuclear Posture Review restored the U.S. commitment to enforce a moratorium on nuclear testing. In his April 5, 2009, speech in Prague, President Obama went further and announced plans to seek ratification of the Comprehensive Test Ban Treaty.

Atomic Islands and "Jellyfish Babies"

Darlene Keju-Johnson (1983)

One important date that I will never forget was in the year 1946. In that year, a Navy official from the U.S. government came to Bikini Island. He told our chief: "We are testing these bombs for the good of mankind, and to end all world wars."

In 1946, very few of us Marshallese spoke English or even understood it. The chief could not understand what it all meant, but there was one word that stuck in his mind and that was "mankind."

The Bikinians were promised that the United States only wanted their islands for a short time. The Bikinians had no choice but to leave their islands. The Navy official did not tell the chief that the Bikinians would never see their home again. Today, Bikini is off-limits for 30,000 years.

The United States put the Bikinians on Rongerik Island and left them there. Rongerik is a sandbar island. There are no resources on it. It was too poor to feed the people. The people of Enewetak Atoll were also relocated. You cannot imagine the psychological problems that people have to go through because of relocation.

In 1954, the United States exploded a hydrogen bomb, code-named BRAVO, on Bikini. It was more than 1,000 times stronger than the Hiroshima bomb. We were never warned that this blast was about to happen. Instead, we experienced white fallout. The southern area of our islands turned yellow. And the children played in it. When the fallout went on their skins, it burned them. People were vomiting. . . .

It was not a mistake. The United States moved the Marshallese in the 1940s when the small bombs were being tested, but when the biggest bomb ever was tested, the Marshallese were not even warned. This is why we believe that we have been used as guinea pigs.

Now we have this problem of what we call "jellyfish babies." These babies are born like jellyfish. They have no eyes. They have no heads. They have no arms. They have no legs. They are not shaped like human beings at all. Some of them have hair on them. And they breathe. They only live for a few hours and, when they die, they are buried right away. A lot of times they don't allow the mother to see this kind of baby because she'll go crazy.

Many women today are frightened of having these "jellyfish babies." I have had two tumors taken out and I fear that if I have children they will be "jellyfish babies" also. These babies are being born not only on the radioactive islands but now throughout the twenty-nine atolls and five islands of the Marshalls. I have interviewed hundreds of Marshallese women. The health problems are on the increase. They have not stopped.

Marshallese are fed up with the Department of Energy and the U.S. government. We are asking for an independent radiological survey. But the United States wouldn't help.

This is what happened with the Rongelap people. They said, "We have had enough! You are not to treat us like animals, like nothing at all! We are moving." So they moved from Rongelap with the Greenpeace ship, *Rainbow Warrior*. The whole island, 350 people, moved to live on Mejato, a small island in the Kwajalein Atoll. By doing this, the Rongelap people said that they don't want to be part of this whole nuclear craziness. Their bottom line is: "We care about our children's future." They had to come to a very hard decision—leaving your island is not easy—but they decided that their children came first. They know they'll be dying out soon, they are dying now—slowly.

We are only a very few thousand people out there on tiny islands, but we are doing our part to stop this nuclear madness. We must come together to save this world for our children and the future generations to come.

[*Darlene Keju-Johnson died of breast cancer at the age of 45.*]

Haunted by Memories of War

S. Brian Willson (2011)

The first target Bao and I visited was just northwest of Highway 4 in southern Vinh Long Province, north of the Bassac River, four or five miles as the crow flies from Binh Thuy, and perhaps eight or nine miles south of Sa Dec. With Bao in the passenger seat of my Jeep, I drove to Can Tho and then took the ferry north across the Bassac River to Highway 4. Within a couple miles we turned onto a side track and soon we saw plumes of smoke behind some high grass. The bombing reportedly had occurred within an hour or so prior to our arrival.

We got out of the Jeep. I began walking cautiously through the high grass, toward the smoke. I heard low moans, then increasingly loud roars.

I looked to my right and saw a water buffalo, a third of its skull gone and a three-foot gash in its belly. I couldn't believe it was still alive. I felt sick to my stomach.

Out of the corner of my eye, perhaps eighty feet toward center-left, I saw one young girl trying to get up on her feet, using a stick as a makeshift cane, but she quickly fell down. A few other people were moving ever so slightly as they cried and moaned on the ground. Most of the human victims I saw were women and children, the vast majority lying motionless. Most, I am sure, were dead. Napalm had blackened their bodies, many of them nearly unrecognizable.

My first thought was that I was witnessing an egregious, horrendous mistake. The "target" was no more than a small fishing and rice farming community—smaller than a baseball playing field. As with most settlements, this one was undefended—we saw no antiaircraft guns, no visible small arms, no defenders of any kind. The pilots who bombed this small hamlet flew low-flying planes, probably A37Bs, and were able to get close to the ground without fear of being shot down, thus increasing the accuracy of their strafing and bombing. They certainly would have been able to see the inhabitants, mostly women with children taking care of various farming and domestic chores. They had the option to abort their mission based on their own observations from the air. They could have been back, safe and sound in Binh Thuy, in less than fifteen minutes. Instead, they had come back with a big thumbs-up, mission accomplished.

Colonel Anh need not have worried about any of his pilots missing their targets. The buildings were virtually flattened by explosions or destroyed by fire. I didn't see one person standing. Most were ripped apart from bomb shrapnel and machine-gun wounds, many blackened by napalm beyond recognition; the majority were obviously children.

I began sobbing and gagging. I couldn't fathom what I was seeing, smelling, thinking. I took a few faltering steps to my left, only to find my way blocked by the body of a young woman lying at my feet. She had been clutching three small, partially blackened children when she apparently collapsed. I bent down for a closer look and stared, aghast, at the woman's open eyes. The children were motionless, blackened blood drying on their bullet- and shrapnel-riddled bodies. Napalm had melted much of the woman's face, including her eyelids, but as I focused on her face, it seemed that her eyes were staring at me.

She was not alive. But at the moment her eyes met mine, it felt like a lightning bolt jolted through my entire being. Over the years, I have thought of her so often I have given her the name "Mai Ly." (I simply rearranged the letters of My Lai, location of the infamous 1968 massacre of Vietnamese villagers by U.S. forces.)

I was startled when Bao, who was several feet to my right, asked why I was crying. I remember struggling to answer. The words that came out astonished me. "She is my family," I said, or something to that effect. I don't know where those words came from. I wasn't thinking rationally. But I felt, in my body, that she and I were one.

Bao just smirked and said something about how satisfied he was with the bombing success in killing communists. I did not reply. I had nothing to say. From that moment on, nothing would ever be the same for me.

I was experiencing such a shock that it did not occur to me to seek medical help for those who still might be alive in the village. We just walked away from the moaning of those still alive. They did not count as human beings. Though I was crying inside, I did nothing. I soon learned that few people survive the suffocation and burns that result from napalm bombing.

The war did not stop to allow me time to assimilate the tragic deaths of young women, children, and old men. . . .

By fate, part of my job was to document bombing casualties in Vinh Long Province. If those numbers were multiplied across the country, I knew the total number of dead and wounded must be enormous. During that same week, Bao and I went to four other bombing "targets." They, too, were inhabited hamlets. Though the week remains a blur, I estimated that we documented somewhere between seven and nine hundred murders of Vietnamese peasants, all due to low-flying fighter-bombers who could see exactly who and what they were bombing.

A consensus has emerged that, from the official invasion of Da Nang on March 8, 1965, through the cease-fire on January 27, 1973—a period of nearly ninety-five months—more than six million Southeast Asians were killed (or, as I would now say, murdered). Over 2,100 people were being killed each day—88 every hour.

I could not talk about this experience for twelve years, and the thought of it still creates tremors in my body. I often find myself crying at the thought of it and, at times, I feel a rage that nearly chokes me.

I now knew, viscerally, the evil nature of the war. I had no choice—God help me!—but to admit that my own country was engaged in an effort that was criminal and immoral beyond comprehension.

As I looked at the blackened, mangled, and maimed bodies of women and children—innocents who had been destroyed by U.S. napalm—I began to see the war in a new way. I suddenly remembered the time when I had tortured my dog, Lucky. The land of my birth, the land of "liberty," was in the grip of the very same monster I had once discovered resided within me. Our arrogant way of life, the American Way of Life, was itself AWOL from the world of universal humanity.

After Vietnam, I knew that my own government was not only criminal but also psychotic. Buried deeper inside me, however, was an even more radical epiphany, the truth Mai Ly offered me through her open eyes. *She is my family.* It would take me many years to understand the real meaning of this experience—that we are all one—a lesson that continues to deepen and expand as I grow older.

The War Zone That Became a New Eden

Lisa Brady (2012)

A thin green ribbon threads its way across the Korean Peninsula. Viewed from space, via composite satellite images, the winding swath clearly demarcates the political boundary between the Republic of Korea (ROK) and the Democratic People's Republic of Korea (DPRK). Its visual impact is especially strong in the west, where it separates the gray, concrete sprawl of Seoul from the brown, deforested wastes south of Kaesong. In the east, it merges with the greener landscapes of the Taebaek Mountain Range and all but disappears.

From the ground, the narrow verdant band manifests as an impenetrable barrier of overgrown vegetation enclosed by layers of fences topped by menacing concertina wire and dotted with observation posts manned by heavily armed soldiers. That a place so steeped in violence still teems with life seems unimaginable. And yet, the Demilitarized Zone, or DMZ, is home to thousands of species that are extinct or endangered elsewhere on the peninsula. It is the last haven for many of these plants and animals and the center of attention for those intent on preserving Korea's rich ecological heritage.

Once known as the "land of embroidered rivers and mountains," the Korean Peninsula has experienced almost continual conflict for over one hundred years, resulting in a severely degraded natural environment. International competition for control over the peninsula's resources left Korea in a precarious position at the start of the twentieth century. The Japanese occupation between 1905 and 1945 brought with it radically increased exploitation of mineral and other resources, resulting in massive deforestation, pollution, and general environmental decline.

Since at least the 1940s, deforestation for fuel wood and clearing for agricultural land has caused significant erosion of the area's mountains and hills and contributed to the siltation of its rivers, streams, and lakes. The 1950–53 war ranged across the entire peninsula, subjecting it to widespread devastation that destroyed cities, roads, forests, and even mountains. In the 1960s and 1970s, unchecked industrialization further undermined the peninsula's ecological health, causing air, water, and soil pollution.

The relative health of the DMZ now stands in stark contrast to the failing ecosystems in both North and South Korea.

Created in 1953 during tense armistice negotiations, Korea's DMZ is at once one of the most dangerous places on Earth and one of the safest. For humans, its thousands of land mines and the millions of soldiers arrayed along its edges pose an imminent threat. But the same forces that prevent humans from moving within the nearly 400 square miles of the DMZ encourage other species to thrive. Manchurian or red-crowned cranes and white-naped cranes are among the DMZ's most famous and visible denizens. Nearly 100 species of fish, perhaps 45 types of amphibians and reptiles, and over 1,000 different insect species are also supposed to exist in the protected zone.

Scientists estimate that more than 1,600 types of vascular plants and more than 300 species of mushrooms, fungi, and lichen are thriving in the

DMZ. Mammals such as the rare Amur goral, Asiatic black bear, musk deer, and spotted seal inhabit the DMZ's land and marine ecosystems. There are even reports of tigers, believed extinct on the peninsula since before Japanese occupation, roaming the DMZ's mountains.

Much of the biodiversity in the DMZ is speculative, extrapolated from spotty scientific studies conducted in the Civilian Control Zone that forms an additional protective barrier along the DMZ's southern edge. Approximate though these studies are, the DMZ's ecological promise is great enough to spur many people to action.

The simmering tensions on the Korean Peninsula over the past sixty-plus years have at times reached a boiling point—from the 1974 discovery of several military incursion tunnels running under the DMZ from North Korea to the 2010 shelling of Yeonpyeong Island—putting the fate of the DMZ and its biodiversity into question. Efforts to protect the DMZ's ecological and cultural offerings have become imperative.

One potential development on the minds of many Koreans is the return of a single Korean nation. A unified Korea would obviate the need for the DMZ and potentially imperil the existence of the various ecosystems the dividing line presently supports.

Many individuals, both on and off the peninsula, hope to preempt the destruction of the DMZ's ecological treasury, should Korea reunify, by establishing a permanently protected area commensurate with the current DMZ boundaries. Solutions already proffered include the creation of a series of dedicated conservation areas in and along the DMZ, the development of ecotourism and educational zones, and attaining World Heritage Site designation through UNESCO.

The Return of the Cranes

Two projects have already proven highly successful: the establishment of crane conservation areas in the DMZ along the Han River estuary and in the Anbyon Plain north of the DMZ near Wonsan, DPRK. The former zone spans the western edge of the DMZ south to Seoul and serves as a safe winter resting area for over 1,000 white-naped cranes. The South Korean government designated the area a national monument in 1976, after scientific studies by George Archibald of the International Crane Foundation and Kim Hon Kyu of Ehwa Women's University determined the region's importance to the continued survival of the species.

The second conservation area, located in the Anbyon Plain, supplements the migratory wintering grounds for both white-naped and red-crowned cranes inside the DMZ in the Cheorwon Basin and along the peninsula's western coast. Since its inception in the 1990s, the Anbyon Plain project has been an international, cooperative effort, bringing in researchers from Korea University in Tokyo, the State Academy of Sciences in Pyongyang, and Tokyo University.

By working in partnership with local farmers to restore riparian, grassland, and forest habitats, the project demonstrates that cranes and humans can thrive in cooperation. Thus far, the project has garnered approval from the North Korean government and has set the stage for scientists and farmers to create both ecologically and economically sustainable agricultural practices that benefit both cranes and people.

Ecotourism and eco-education are also making headway. South Korea's official tourism site offers the opportunity to visit an area south of the DMZ that it calls the Peace-Life Zone. The guidebook for the tour incorporates extensive photographs highlighting the natural beauty of the region and advertises the DMZ as "a peaceful place" that "marks the last untouched Cold War border in the world today."

Marketing and propaganda aside, the tour is evidence of the ROK's willingness to promote and support efforts to preserve the area's ecological sustainability. Similarly, in early 2010, the Catholic University of Korea and SungKongHoe University developed a joint course for students called "Life and Peace of the DMZ." The course examined the history of the Korean War, media representations of the DMZ, and the sociology and ecology of the area.

Perhaps the most comprehensive plans involve efforts to permanently set aside the DMZ and adjacent areas as an ecological preserve and cultural site. Several international and Korean NGOs are involved in disseminating information, gathering public support, and working with government agencies on both sides of the divide to achieve official recognition of the DMZ's importance both environmentally and historically and subsequent protection of the area as an international transboundary peace park.

Among the leading associations, and one of the first organized for the purpose, is the DMZ Forum, whose mission is "to support conservation of the unique biological and cultural resources of Korea's Demilitarized Zone, transforming it from a symbol of war and separation to a place of peace among humans and between humans and nature." The Forum's primary and long-term goal is to gain approval from both sides to nominate the entire DMZ and its adjacent zones as a cultural and natural World Heritage Site. This transboundary park would incorporate marine ecosystems in the West Sea, the Han River estuary and its lowland plains, the high mountains of the Taebaek Mountain Range, and the East Sea tidal areas, providing a nearly comprehensive representation of the peninsula's ecological systems. Combined, the area would incorporate over 400 square miles of protected ecosystems.

If the DMZ has served to remind the Korean people and the world of the lasting legacy of ideological and military conflict, then it has also become a symbol for a new generation wanting to preserve a reinvigorated Korean ecology. The war was a human tragedy, but out of terrible loss may come the prospect for ecological health and mutual cooperation. Korea's political past and its environmental future are inextricably linked. The DMZ has become a green ribbon of hope, representing Korea's promise for a healthy, peaceful future.

Part III

ECOLIBRIUM—PATHWAYS *to a* PLANET *at* PEACE

History has witnessed the failure of many endeavors to impose peace by war, cooperation by coercion, unanimity by slaughtering dissidents. . . . A lasting order cannot be established by bayonets.

—Ludwig von Mises, Austrian-American economist and author

I hate war as only a soldier who has lived it can, only as one who has seen its brutality, its futility, its stupidity. . . . I think that people want peace so much that one of these days governments had better get out of the way and let them have it.

—Dwight D. Eisenhower

Religion, custom, and law are supposed to guide human society in the search for security and justice, but they have proven to be flawed in this regard. While many religious leaders have spoken out eloquently against war, in practice, most world religions have done little to stop it, essentially giving silent approval and letting it rage under certain—often very elastic—circumstances.

The Old Testament's biblical commandment "Thou shalt not kill" seems an immutable call to pacifism and nonviolence—a moral foundation upon which civilized humanity can build. But, for religious adherents over several centuries, the Sixth Commandment has generally not been a hard-and-fast law. Not even Moses, the custodian of the Stone Tablets, appeared to be bound by its strictures. According to Exodus 32:28, the first thing Moses does, after descending from Mount Sinai to share the Decalogue with his followers, is to order the Levites to murder 3,000 of their companions for worshiping the Golden Calf (a violation of the Second Commandment).

The New Testament features a different case of divine wrath—this time directed against nature. In Mark 11:13–14, a hungry Jesus curses a fig tree when he discovers it has no fruit. The tree withers and dies. In contrast, we have the Prophet Mohammad's instruction that believers plant a tree, even if the world is ending.

While the Torah and the Hebrew Bible forbid murder (*ratzah*)—the shedding of innocent blood—they allow the killing of enemy fighters during war (*harag*). The Koran (17:33) makes a similar distinction: "Nor take life—which Allah has made sacred—except for just cause."

The phrase "just war"—a holy loophole introduced in Saint Augustine's *City of God* in AD 426—was later expanded by Thomas Aquinas (1225–1274). From 1095 to 1291, the Roman Catholic Church promoted a series of Holy Crusades as divinely sanctioned invasions—incursions that cost, by some estimates, 1.7 million lives.

International treaties to restrain or abolish war have a long history, but it's a history of lofty goals undone by compromise. The 1907 Hague

Convention and Protocol I to the Geneva Conventions failed to prevent the ghastly conduct that rained bombs and poison gas on the battlefields of Europe. The 1907 "Laws and Customs of War on Land" and "Bombardment by Naval Forces in Time of War" contained no prohibitions against aerial bombardment. Under current humanitarian law, strategic bombers, attack fighters, and drones can all be used as long as they observe "military necessity," "distinction" and "proportionality." This leaves armies free to attack "military objectives" and cause harm to civilian property or lives so long as the damage is "proportional."

Following World War I, which featured the novelty of German Zeppelins dropping bombs on Britain and France, world powers drafted The Hague Rules of Air Warfare. Unfortunately, the 1923 treaty was dismissed as "unrealistic" and never adopted.

The horrendous aerial bombings of the Spanish Civil War (1936–39) and the Second Sino-Japanese War (1936–38) prompted the League of Nations to craft a convention calling for the protection of civilians during aerial bombardments, but this 1938 convention was never ratified.

With the outbreak of World War II, Britain, Germany, and the U.S. agreed not to bomb civilian targets but, as thousands of historic photos attest, this pledge was not honored.

Following the end of World War II, the Nuremberg Charter condemned as a "war crime" the "wanton destruction of cities, towns, or villages, or destruction not justified by military necessity." During the Nuremberg and Tokyo war crimes trials, however, these specific war crimes were not addressed.

Beginning in 1949, a series of treaties governing the laws of war were adopted under the Geneva Conventions. Article 35(3) specifies that: "It is prohibited to employ methods or means of warfare which are intended, or may be expected, to cause widespread, long-term and severe damage to the natural environment." The Convention's Article 55 prohibits reprisals against the natural environment and imposes a duty of care on warring parties.

The Fourth Convention specifically addressed civilian rights in occupied territories but, once again, failed to address the issue of aerial bombardment. It wasn't until 1977 that Protocol I was attached to the Geneva Conventions to specifically ban deliberate or indiscriminate attacks on civilians.

In a rare 1963 decision, a Japanese court ruled in Shimoda v. the State that the extraordinary power of the atom bomb and the distance from legitimate military land targets made the U.S. atomic attacks on Hiroshima and Nagasaki "an illegal act of hostilities under international law." In 1996, the International Court of Justice concluded that the indiscriminate damage caused by nuclear weapons made their use illegal under international law. In 2007, this decision was adopted by the non-state International Peoples' Tribunal.

These rulings did not sit well with Washington. Following the adoption of the Rome Statute establishing the International Criminal Court on July 19, 1998, the United States became one of only seven countries voting against the ICC. In 2001, Undersecretary of State for Arms Control and International Security Affairs John R. Bolton explained why the United States would not ratify the ICC Treaty. Bolton bristled at the idea that the ICC might dare to "find the United States guilty of a war crime for dropping atomic bombs on Hiroshima and Nagasaki. This is intolerable and unacceptable." (Upon receiving the required sixty ratifications, the Rome Statute entered into force on July 1, 2002.)

Nearly every "war crime" statute crafted in Geneva, Nuremberg, or The Hague contains an "escape clause." Murdering innocent civilians, destroying cultural artifacts, poisoning waterways, incinerating forests are all permissible if "justified by military necessity." In 1996, the International Court of Justice ruled that it could not conclude "definitively whether the threat or use of nuclear weapons would be lawful or unlawful in an extreme circumstance of self-defense."

While international bodies continue to struggle with rules for "proper" military conduct and definitions of war crimes, civil society continues to push for grassroots and institutional solutions.

As international law professor Steven Freeland, author of *Addressing the Intentional Destruction of the Environment during Warfare*, points out: "Just as international law has made great strides forward by classifying rape during armed conflict as a war crime—a crime against humanity, or even genocide in certain circumstances—we should recognize that intentional environmental destruction can also constitute an international crime."

While human nature and human behavior are largely responsible for the continued aggravations of war, our habits and actions are subject to change. We can chose to act in defense of our rivers, oceans, forests, and skies. We have the power to choose life. We can dedicate ourselves to putting the Earth back in balance. Others already have embarked on the path to Ecolibrium. All we need do is to follow their signposts—or find our own trails.

Mapping the Terrain

Take the Profit Out of War is Major Gen. Smedley Butler's simple prescription for putting an end to unprovoked and increasingly endless conflicts. Butler's experience as one of America's most decorated soldiers underscores the tactical wisdom of his 1935 proposal to reject the "racket" of waging (and staging) wars.

Ecology and War contains the text of a historic appeal written by David Brower, a legendary environmental leader who earned a Bronze Star in World War II and returned home to defend the wilderness as head of the Sierra Club, Friends of the Earth, and Earth Island Institute. Brower was one of the first to condemn the environmental impacts of war.

Why We Oppose War and Militarism is a collective statement of concern signed by a global coalition of more than 100 peace and social justice organizations that was issued in advance of the 2003 U.S. invasion of Iraq.

In Defense of the Environment, a 2003 speech by former United Nations Environment Programme director Klaus Toepfer (1998–2006), underscores how words alone have failed to move the world away from war. He calls for more decisive action and, in particular, laws that safeguard peace by protecting the environment.

Protecting Nature from War, a UNEP policy statement from 2009, marks a significant attempt to extend the rule of law to the protection of nature. The next step is to recognize the defiling of the natural world as an international war crime—without exceptions.

Strategies for a More Peaceful World reviews a growing number of civil society campaigns intended to address the devastation of global militarism. Our short list documents more than thirty local and global initiatives and provides contact information for more than fifty leading international peace organizations.

TOWARD ECOLIBRIUM

Take the Profit Out of War

Major-General Smedley Butler (1935)

It's a racket, all right. A few profit—and the many pay. But there is a way to stop it. You can't end it by disarmament conferences. You can't eliminate it by peace parleys at Geneva. Well-meaning but impractical groups can't wipe it out by resolutions. It can be smashed effectively only by taking the profit out of war.

The only way to smash this racket is to conscript capital and industry and labor before the nation's manhood can be conscripted. One month before the Government can conscript the young men of the nation—it must conscript capital and industry and labor. Let the officers and the directors and the high-powered executives of our armament factories and our munitions makers and our shipbuilders and our airplane builders and the manufacturers of all the other things that provide profit in wartime as well as the bankers and the speculators, be conscripted—to get $30 a month, the same wage as the lads in the trenches get. [That soldier's wage would be $525.56 in 2016 dollars. The monthly wage for a modern E-1 (enlisted soldier) is $1599.90.]

Let the workers in these plants get the same wages—all the workers, all presidents, all executives, all directors, all managers, all bankers—yes, and all generals and all admirals and all officers and all politicians and all government officeholders—everyone in the nation be restricted to a total monthly income not to exceed that paid to the soldier in the trenches!

Let all these kings and tycoons and masters of business and all those workers in industry and all our senators and governors and majors pay half of their monthly $30 wage to their families and pay war risk insurance and buy Liberty Bonds.

Why shouldn't they?

They aren't running any risk of being killed or of having their bodies mangled or their minds shattered. They aren't sleeping in muddy trenches. They aren't hungry. The soldiers are!

Give capital and industry and labor thirty days to think it over and you will find, by that time, there will be no war. That will smash the war racket—that and nothing else.

Maybe I am a little too optimistic. Capital still has some say. So capital won't permit the taking of the profit out of war until the people—those who do the suffering and still pay the price—make up their minds that those they elect to office shall do their bidding, and not that of the profiteers.

Another step necessary in this fight to smash the war racket is the limited plebiscite to determine whether a war should be declared. A plebiscite not of all the voters but merely of those who would be called upon to do the fighting and dying. There wouldn't be very much sense in having a seventy-six-year-old president of a munitions factory or the flat-footed head of an international banking firm or the cross-eyed manager of a uniform manufacturing plant—all of whom see visions of tremendous profits in the event of war—voting on whether the nation should go to war or not. They never would be called upon to shoulder arms—to sleep in a trench and to be shot. Only those who would be called upon to risk their lives for their country should have the privilege of voting to determine whether the nation should go to war.

There is ample precedent for restricting the voting to those affected. Many of our states have restrictions on those permitted to vote. In most, it is necessary to be able to read and write before you may vote. In some, you must own property. It would be a simple matter each year for the men coming of military age to register in their communities as they did in the draft during the World War and be examined physically. Those who could pass and who would therefore be called upon to bear arms in the event of war would be eligible to vote in a limited plebiscite. They should be the ones to have the power to decide—and not a Congress, few of whose members are within the age limit and fewer still of whom are in physical condition to bear arms.

A third step in this business of smashing the war racket is to make certain that our military forces are truly forces for defense only.

At each session of Congress the question of further naval appropriations comes up. The swivel-chair admirals of Washington (and there are always a lot of them) are very adroit lobbyists. And they are smart. They don't shout that "we need a lot of battleships to war on this nation or that nation." Oh no. First of all, they let it be known that America is menaced by a great naval power. Almost any day, these admirals will tell you, the great fleet of this supposed enemy will strike suddenly and annihilate 125,000,000 people. Just like that. Then they begin to cry for a larger Navy. For what? To fight the enemy? Oh my, no. Oh, no. For defense purposes only.

Then, incidentally, they announce maneuvers in the Pacific. For defense. Uh, huh.

The Pacific is a great big ocean. We have a tremendous coastline on the Pacific. Will the maneuvers be off the coast, two or three hundred miles? Oh, no. The maneuvers will be 2,000, yes, perhaps even 3,500 miles, off the coast.

The Japanese, a proud people, of course will be pleased beyond expression to see the United States fleet so close to Nippon's shores. Even as pleased as would be the residents of California were they to dimly discern through the morning mist, the Japanese fleet playing at war games off Los Angeles.

The ships of our navy, it can be seen, should be specifically limited, by law, to within 200 miles of our coastline. Had that been the law in 1898, the *Maine* would never have gone to Havana Harbor. She never would have been blown up. There would have been no war with Spain with its attendant loss of life. Two hundred miles is ample, in the opinion of experts, for defense purposes. Our nation cannot start an offensive war if its ships can't go further than 200 miles from the coastline. Planes might be permitted to go as far as 500 miles from the coast for purposes of reconnaissance. And the army should never leave the territorial limits of our nation.

To summarize: Three steps must be taken to smash the war racket.

We must take the profit out of war.

We must permit the youth of the land who would bear arms to decide whether or not there should be war.

We must limit our military forces to home defense purposes.

Ecology and War

David Brower / Friends of the Earth (1982)
[*Adapted from an open letter to President Richard Nixon in 1970 and a full-page ad addressed to President Ronald Reagan, published in the October 1982 issue of FOE's newsmagazine,* Not Man Apart.]

This advertisement is placed by Friends of the Earth, a conservation organization, but it concerns war.

Until recently, we were content to work for our usual constituency: Life in its miraculous diversity of forms. We have left it for others to argue about war.

Ecology teaches us that everything is irrevocably connected. Whatever affects life in one place—any form of life, including people—affects other life elsewhere.

DDT on American farms finds its way to Antarctic penguins.

Pollution in a trout stream eventually pollutes the ocean.

Smog over London blows over to Sweden.

The movement of a dislodged, hungry, war-torn population affects conditions and life wherever they go.

An A-bomb explosion spreads radiation everywhere.

It is all connected.

Nuclear war would kill life of *all* kinds, indiscriminately, on a scale and for a length of time so great that it qualifies as the major ecological issue of our time. From an ecological point of view, the war has already begun, as the mere preparations for it are causing illness and deprivation to humans and bringing harm to the planet's life-sustaining abilities.

We have been living with the nuclear threat too long. That nations continue to threaten each other does violence to the human spirit. We are told not to think about it, but to leave it to national security planners who talk of "counterforce targeting," "equivalent megatons," "industrial survivability"—phrases that avoid the actuality of millions of dead children, vaporized forests, disappeared cities.

We have heard this kind of language before. Forests are "board feet," rivers are "acre feet," nature itself is a "resource." No sense of awe of the natural world. No connection or caring. It's the same mentality that can plan the death of whole populations. They sit before computers plotting war scenarios and missile trajectories, dazzled by automated weapons and computer targeting. To them it is just blips on a video screen, like some kind of hot game of Missile Command. Except at the end, there are 500 million dead bodies and a world laid waste for eons.

Even if the shooting war never actually starts, every stage in preparation for it brings danger to human beings and the planet. Digging uranium from the ground causes lung cancers among miners. The transit of nuclear materials on roads brings the chance of deadly accidents. Nuclear plants make disaster or terrorism a present danger. Continued testing of weapons releases radiation into soil and air. *All* nuclear production adds to the stockpiles of atomic wastes—some deadly to life for 250,000 years or more.

The *Global 2000 Report to the President* states that catastrophe awaits if nations don't address the following: 1) an imminent worldwide shortage of water, 2) the loss of forest cover, 3) the rapidly expanding deserts and the deterioration of the world's topsoil, affecting the food supply, 4) the ravaging of the world's remaining fuel and minerals, 5) the destruction of animal habitats and, with them, the planet's genetic reserves. These are symptoms of breakdown of the Earth's life-support system—a *holocaust*, except in slow motion, without the fireball.

President Reagan responded to the report by firing its authors. He dismantled U.S. agencies that research these problems. And he has increased the rate at which we devour energy and minerals to feed the military machine.

Both [Reagan and Soviet leader Leonid Brezhnev] see the world in terms of power, domination, "national interest," "spheres of influence," "massive retaliation." They both seek to "negotiate from strength," i.e., by intimidation. Can this possibly succeed? ("You don't increase your security by decreasing the security of your opponent"—Richard Barnett.)

Mr. Reagan says he wants peace, but he budgets more for war than anyone in history, at the expense of the economy, suffering Americans, and the environment. He is exporting more than $25 billion dollars in arms to other nations. He is also increasing exports of nuclear technology, bringing bomb capability to many small countries. He speaks of "freedom" but supports governments that oppress their own people. Mr. Brezhnev seems the same. He speaks of peace but builds arms at the expense of his own people and economy.

The world is sick of watching this!! Two bullies, facing each other down, acting out of pride, machismo, and insincerity. And apparently willing to bring all the rest of us along with them to defend their stuck positions. How do you get through to men like this?

We need a profound change. A new political language that recognizes that real security emerges only when nations befriend each other. Belligerency, competition for resources on a finite planet, exploitation of people and nature are behaviors that have failed. They are out of date. We need to seek people skilled in the arts of peace, accommodation, compromise. Not shy to speak of love for all creatures and for whom war is a violation of natural law, a breakdown of the civilized experiment. We need to articulate a system of values that reflects this.

Why We Oppose War and Militarism

On February 20, 2003, Environmentalists Against War (a global coalition of antiwar organizations) held a press conference at the San Francisco office of the Sierra Club to announce its opposition to the George W. Bush administration's imminent attack on Iraq. A joint statement—signed by more than 100 environmental, antiwar, and social-justice organizations from around the world—enumerated 11 reasons to reject war and militarism. The declaration read as follows:

1. War Kills People

War is humankind's deadliest activity. From 500 BC to AD 2000 there have been 1,022 major documented wars. Between AD 1100 and 1925, around 35.5 million died in European wars alone. In the twentieth century, an estimated 165 wars were responsible for the deaths of 165 million to 258 million. Military conflicts caused the deaths of as many as 6.25 percent of all the people born during the twentieth century. Approximately 8.4 million soldiers and 5 million civilians died in World War I. World War II claimed the lives of 17 million soldiers and 34 million civilians. Seventy-five percent of those killed in modern war are civilians. War disproportionately kills and injures women, children, the elderly, minorities, and the poor.

2. War Destroys Nature

War destroys wildlife, disrupts native habitats, and contaminates the land, air, and water. The damage can last for generations. The United States dropped 25 million bombs and 19 million gallons of Agent Orange herbicide and other chemical weapons on the forests, fields, and farms of Vietnam. Millions of acres from Russia's Baltic Sea to the Pacific Ocean have been contaminated by military chemicals and radioactive wastes. In Cambodia, 1,300 square miles are salted with several million mines that continue to kill wildlife and humans. Angola's environment is burdened with more than 10 million land mines. Cluster bombs, thermobaric explosions, chemical and biological weapons, and projectiles made with radioactive depleted uranium are indiscriminate weapons of mass destruction.

3. War Devastates Society

War destroys villages, farmland, and urban infrastructure. Wars destroy irreplaceable cultural artifacts, ancient landmarks, and archeological sites. The United States dropped 88,000 tons of bombs on Iraq in 1991, destroying 9,000 homes, water systems, power plants, critical bridges, and four major dams. The resulting health emergency contributed to the deaths of 500,000 Iraqi children. In 2002, the U.S. dropped a quarter-million cluster bomblets on Afghanistan. In 2003, the U.S. dropped 28,000 rockets, bombs, and missiles on Iraq. In the past twenty-five years, war has devastated cities and villages around the world, leaving lasting damage in such diverse countries as Sudan, El Salvador, Mozambique, Angola, Lebanon, Yugoslavia, Rwanda, Afghanistan, Liberia, Uganda, Colombia, Somalia, Congo, Iraq, Burundi, Iran, and Ethiopia.

4. War Consumes Resources

A vast global military empire must be maintained to feed the world's oil-based economies. Waging war requires burning vast stores of oil and generates significant spikes of greenhouse gasses. World War II consumed 6 to 9 billion barrels of oil. Desert Storm: 45 million barrels. The Pentagon consumed 134 million barrels in 2001. The world's armies consume nearly two billion barrels of oil annually. The Pentagon is the largest consumer of oil, chemicals, precious metals, paper, and wood.

5. War Pollutes

Bombs, missiles, shells, bullets, and military fuels poison our land, air, and water with lead, nitrates, nitrites, hydrocarbons, phosphoros, radioactive debris, corrosive and toxic heavy metals. Unexploded ordnance lies scattered over more than 15 million U.S. acres. The world's armies are responsible for as much as 10 percent of global air pollution. The 1991 Gulf War generated 80,000 tons of global-warming gases. On any given day, more than 60,000 U.S. troops are engaged in operations or military exercises in about 100 foreign countries. The Pentagon is the world's largest polluter, generating 750,000 tons of hazardous wastes each year. U.S. military bases have polluted communities in Canada, Germany, Great Britain, Greenland, Iceland, Italy, Panama, the Philippines, South Korea, Spain, and Turkey. There are more than 14,000 contaminated military sites in the United States, many located near low-income neighborhoods and communities of color.

6. War Is Costly

The cost of all U.S. military conflicts from the Revolutionary War to World War II has been estimated at more than $4 trillion. Increased military spending drains funds from critical social, educational, medical, and environmental needs. In the United States, 51 percent of the 2003 discretionary federal budget went to the military. Global military spending hit $798 billion in 2000. Global spending on the military now stands at around $842 billion a year. It costs $2.2 billion to build, support, and operate one naval battle group for one year. Thirteen million dollars could provide access to clean water for 80,000 Third World villages. The cost of one $1.5 billion Trident submarine could immunize the world's children against six deadly diseases and prevent 1 million deaths a year. CNN observed on March 20, 2003: "The cost of the first twenty-five Tomahawk Missiles launched in the first hour of the first day in the war with Iraq was more than fifty times the annual HUD budget to End Homelessness in America."

7. Militarism Undermines Peace

War diverts vast amounts of capital resources and human energy from serving critical social, educational, medical, and environmental needs into efforts that are destructive and deadly. Unsustainable economies must rely on the use of military force to secure control of essential foreign resources—oil, uranium, and metals. In 2001, 247,000 U.S. soldiers were stationed at 752 bases in more than 130 countries. Militarily dominant states are prone to acts of aggression—and aggression invites retaliation. The United States is the world's largest supplier of weapons ($31.8 billion in 2000). Many countries that buy U.S. weapons are repressive regimes that ignore the needs of their own citizens. Around the world, militarism impoverishes the many and enriches the few. The only beneficiaries from this dangerous instability are the world's weapons manufacturers and war profiteers.

8. Militarism Weakens Democracy

Military organizations are inherently authoritarian systems that promote a cult of obedience rather than a culture of independence. Since

1859, U.S. troops have intervened militarily around the world more than 160 times—an average of once a year. To justify these interventions, U.S. officials have lied to the American people about the pretexts underlying the wars. Around the world, declarations of war and martial law—frequently based on misrepresented or staged provocations—have been used to institute press censorship, curtail dissent, and imprison political opponents. Militarization and the war on terrorism have been used as an excuse to erode political and civil liberties. Under the U.S. PATRIOT Act, environmental protests now can be defined as terrorist acts. Around the world, the military insists on being exempt from environmental and civil laws.

9. Militarism Distorts Science

Militarism encourages the development of ever-deadlier weapons. Universities and corporations that could be devoting time, talent, and resources to addressing problems of poverty, sickness, and injustice are instead designing exotic new military technologies. These exotic weapons include: chemical weapons, ethnically targeted weapons, electromagnetic guns, mind-altering drugs, miniaturized surveillance technology, and "less-than-lethal" weapons to be used to control a country's own citizens. The United States spends more than $58 billion a year on military research and development. Worldwide, more than 50 million scientists, researchers, and workers are employed in the arms industry.

10. Militarism Promotes Racism

Militarism requires citizens of one country to believe that the citizens of competing nations are intrinsically evil or even subhuman. Military bases, weapons depots, storage yards, and military exercises expose poor neighboring communities to debilitating levels of noise, chronic air pollution, chemical contamination, and the risk of accidental death or injury. Nuclear ore is extracted from native lands, nuclear weapons are tested on native lands, and nuclear wastes are deposited on native lands.

11. Militarism Threatens Human Survival

The United States has threatened other countries with the preemptive use of nuclear weapons—the ultimate weapons of mass destruction. U.S. nuclear attacks against the civilian populations of Hiroshima and Nagasaki killed 210,000, while blast survivors were doomed to slow, lingering deaths. Fallout from open-air nuclear testing is expected to eventually kill about 2.4 million people worldwide. Nuclear weapons stockpiled by Israel, India, Pakistan, Russia, China, and Britain have the potential to end human civilization. An exchange of nuclear weapons between India and Pakistan could kill 30 million. These costly and dangerous stockpiles must be dismantled and destroyed.

In Defense of the Environment

Klaus Toepfer, United Nations Environment Programme (2003)

One can easily clean up the language of war—"collateral damage, friendly fire, smart bombs"—but cleaning up the environmental consequences is a far tougher task.

Undoubtedly it is the loss of human life, the suffering of those made homeless and hungry that must be our primary, first, concern. But all too often, the impact on the Earth's life-support systems is ignored, and ignored, I would suggest, at our peril as the growing expertise of the United Nations Environment Programme's Post Conflict Assessment Branch is suggesting.

Environmental security, both for reducing the threats of war and in successfully rehabilitating a country following conflict, must no longer be viewed as a luxury but needs to be seen as a fundamental part of a long-lasting peace policy.

Few can forget the lakes and pools of petroleum, the TV images of smoke and flames, turning day into night, during the 1991 conflict in Kuwait. An estimated 700 wells were damaged, destroyed, and sabotaged, triggering pollution of water supplies and the seas, the impact of which is still being felt.

It has been suggested that, as a result of the soot, death rates in Kuwait rose by 10 percent over the following year. The only good news was that the over four million tons of soot and sulfur did not climb higher than 5,000 meters (16,400 feet), otherwise there could have been potentially severe dangers to the regional and possibly global climate.

There are many indirect impacts of war on the environment, too. The International Campaign to Ban Landmines, which helped inspire an international convention, says that tens of millions of explosives remain scattered around the world in former conflict areas like Afghanistan, Cambodia, Bosnia, and on the African continent.

These are not only horrific hazards for people—maiming and killing returning refugees and local villagers—but they also effectively bar people from productive land, forcing them to clear forests and other precious areas for agriculture with consequences for the fertility of soils, accelerated land degradation, and loss of wildlife.

Warring factions and displaced civilian populations can take a heavy toll on natural resources. Decades of civil war in Angola have left its national parks and reserves with only 10 percent of the original wildlife. Sri Lanka's civil war has led to the felling of an estimated five million trees, robbing farmers of income. Many poor people in developing countries critically depend on forests for food and medicines.

Many conflicts on continents like Africa have been driven or at the very least fueled by a greed for minerals such as diamonds and oil or timber. Some individuals and groups can make a fortune under the cloak of an ideologically motivated war. It is estimated that UNITA rebels in Angola made over $4 billion from diamonds between 1992 and 2001. The Khmer Rouge was, by the mid-1990s, making up to $240 million a year from exploiting Cambodia's forests for profit.

As the world's life-support systems and natural resources become increasingly scarce, so the possibility of conflict rises. Water, the most precious resource on Earth and crucial for all life, is not evenly shared across the world and between nations. There are 263 river basins, shared by 145 countries. But just 33 nations have more than 95 percent of these rivers within their territories.

By 2032, half the world's population could be living in severely water-stressed areas. Daily, 6,000, mainly children, die as a result of poor or nonexistent sanitation or for want of clean water. It is the equivalent to a quarter of the population of a large capital city like London dying every year. Unless countries learn to use water wisely, learn to share, there will be instability and there will be tensions of the kind that can precipitate war.

Countering this is sustainable development in action. We have an alliance against terrorism, we need an alliance against poverty and solidarity with the marginalized, and we need to defend nature and our natural resources. For little will ever be achieved in terms of conservation of the environment and natural resources if billions of people have no hope, no chance to care.

So we need, above all, environment policy as a precautionary peace policy. Governments are also waking up to the need to rehabilitate the environment if all else fails and conflict occurs. Many are now recognizing that a polluted environment, that contaminated water supplies and sullied land and air, are not a long-term recipe for stability.

There is endless debate before and after a war about the economic costs including the costs of bombs and the costs of humanitarian relief. We need to cost the environmental cleanup, too.

We have the Geneva Conventions, aimed at safeguarding the rights of prisoners and civilians. We need similar safeguards for the environment. Every effort must be made to limit the environmental destruction. Using the environment as a weapon must be universally condemned, must be denounced as an international crime against humankind, against nature.

Protecting Nature from War

United Nations Environment Programme (2009)

The environment continues to be the silent victim of armed conflicts worldwide. The toll of warfare today reaches far beyond human suffering, displacement, and damage to homes and infrastructure. Modern conflicts also cause extensive destruction and degradation of the environment. Environmental damage, which often extends beyond the borders of conflict-affected countries, can threaten the lives and livelihoods of people well after peace agreements are signed.

The existing international legal framework contains many provisions that either directly or indirectly protect the environment and govern the use of natural resources during armed conflict. In practice, however, these provisions have not always been effectively implemented or enforced . . . with one notable exception: holding Iraq accountable for damages caused during the 1990–91 Gulf War, including billions of dollars' worth of compensation for environmental damage.

Public concern regarding the targeting and use of the environment during wartime first peaked during the Vietnam War. The use of the toxic herbicide Agent Orange, and the resulting massive deforestation and chemical contamination it caused, sparked an international outcry leading to the creation of two new international legal instruments. The Environmental Modification Convention was adopted in 1976 to prohibit the use of environmental modification techniques as a means of warfare. Additional Protocol I to the Geneva Conventions, adopted in the following year, included two articles (35 and 55) prohibiting warfare that may cause "widespread, long-term and severe damage to the natural environment."

The adequacy of these two instruments, however, was called into question during the 1990–91 Gulf War. The extensive pollution caused by the intentional destruction of over 600 oil wells in Kuwait by the retreating Iraqi army and the subsequent claims for $85 billion in environmental damages led to further calls to strengthen legal protection of the environment during armed conflict.

In 1992, the UN General Assembly held an important debate on the protection of the environment in times of armed conflict. While it did not call for a new convention, the resulting resolution (RES 47/37) urged Member States to take all measures to ensure compliance with existing international law on the protection of the environment during armed conflict. It also recommended that States take steps to incorporate the relevant provisions of international law into their military manuals and ensure that they are effectively disseminated.

As an outcome of the UN debate, the International Committee of the Red Cross (ICRC) issued a set of guidelines in 1994 that summarized the existing applicable international rules for protecting the environment during armed conflict. These guidelines were meant to be reflected in military manuals and national legislation. Yet armed conflicts have continued to cause significant damage to the environment—directly, indirectly, and as a result of a lack of governance and institutional collapse.

For instance, dozens of industrial sites were bombed during the Kosovo conflict in 1999, leading to toxic chemical contamination at several hotspots. In another example, an estimated 12,000 to 15,000 tons of fuel oil were released into the Mediterranean Sea following the bombing of the Jiyeh power station during the conflict between Israel and Lebanon in 2006.

In recent years, concern has also been raised about the role of natural resources—particularly "high-value" resources—in generating revenue for financing armed forces and the acquisition of weapons. Indeed, easily captured and exploitable resources often prolong and alter the dynamics of conflict, transforming war into an economic rather than purely political activity. Since 1990, at least eighteen civil wars have been fueled by natural resources: diamonds, timber, oil, minerals, and cocoa have been exploited in internal conflicts in the Democratic Republic of the Congo, Côte d'Ivoire, Liberia, Sierra Leone, Angola, Somalia, Sudan, Indonesia, and Cambodia.

Given that natural resources such as water, soil, trees, and wildlife are the "wealth of the poor," their damage and destruction during armed conflict can undermine livelihoods, act as a driver of poverty and forced migration, and even trigger local conflict. As a result, successful peacebuilding—from reestablishing safety, security, and basic services to core government functions and the economy—fundamentally depends on the natural resource base and its governance structure. Natural resources themselves can either unite or divide post-conflict countries depending on how they are managed and restored. It is thus paramount that they be protected from damage, degradation, and destruction during armed conflict.

Findings from More Than Twenty UNEP Post-Conflict Studies, 1999–2009

- Articles 35 and 55 of Additional Protocol I to the 1949 Geneva Conventions do not effectively protect the environment during armed conflict due to the stringent and imprecise threshold required to demonstrate damage: while these two articles prohibit "widespread, long-term and severe" damage to the environment, all three conditions must be proven for a violation to occur.
- To improve the effectiveness of Articles 35 and 55 of Additional Protocol I, clear definitions are needed for "widespread," "long-term," and "severe." As a starting point, the precedents set by the 1976 ENMOD convention should serve as the minimum basis, namely that "widespread" encompasses an area on the scale of several hundred square kilometers; "long-term" is for a period of months, or approximately a season; and "severe" involves serious or significant disruption or harm to human life, natural economic resources, or other assets.
- The majority of international legal provisions protecting the environment during armed conflict were designed for international armed conflicts and do not necessarily apply to internal conflicts: given that most armed conflicts today are non-international or civil wars, much of the existing legal framework does not necessarily apply.
- The provisions for protecting the environment during conflict under the four bodies of international law have not yet been seriously applied in international or national jurisdictions. Only a very limited number of cases have been brought before national, regional, and international courts and tribunals.
- Destruction of the environment and depletion of natural resources may be a material element or underlying act of other crimes contained within the Rome Statute. It is therefore subject to criminal liability and prosecution by the International Criminal Court. Acts of pillage as a war crime are of particular

interest and could be used to prosecute the practice of looting natural resources during conflicts.

- Linking environmental damage to the violation of fundamental human rights offers a new way to investigate and sanction environmental damages, particularly in the context of non-international armed conflicts. A variety of human rights fact-finding missions, including that led by Judge Richard Goldstone in the Gaza Strip in 2009, have investigated environmental damages that have contributed to human rights violations.
- To ensure that environmental violations committed during warfare are prosecuted, provisions of international law that protect the environment in times of conflict should be fully reflected at the national level.
- The UN Compensation Commission was established by the Security Council to process compensation claims relating to the 1990–91 Gulf War. United Nations Member States may want to consider how a similar structure could be established as a permanent body—either under the General Assembly or under the Security Council—to investigate and decide on alleged violations of international law during international and non-international armed conflict.
- A new legal instrument granting place-based protection for critical natural resources and areas of ecological importance during international and non-international armed conflicts should be developed. This could include protection for watersheds, groundwater aquifers, agricultural and grazing lands, parks, national forests, and the habitat of endangered species. At the outset of any conflict, critical natural resources and areas of ecological importance would be delineated and designated as "demilitarized zones." Parties to the conflict would be prohibited from conducting military operations within their boundaries.
- The UN General Assembly should consider requesting the secretary-general to submit a report annually on November

6 (International Day for Preventing the Exploitation of the Environment in War and Armed Conflict) on the environmental impacts of armed conflicts. The report should detail the direct, indirect, and institutional environmental impacts caused by ongoing and new international and non-international armed conflicts. The report should also recommend how the environmental threats to human life, health, and security can be addressed and how natural resources and the environment in each can be used to support recovery and peacebuilding.

Children can move the Earth. PHOTO BEN KERCKX (PIXABAY).

STRATEGIES FOR A MORE PEACEFUL WORLD

The *Global Peace Index*—produced annually by the Institute for Economics and Peace—measures the state of peace in 162 countries according to 23 indicators that gauge the absence or the presence of violence. The 2016 *Index* (http://economicsandpeace.org) described a world that was less peaceful in 2015 than it was in 2008—a world increasingly divided between countries enjoying unprecedented levels of peace and prosperity while others spiraled further into violence and conflict. Eighty-one countries had become more peaceful, while seventy-nine had experienced conflicts. Europe, the world's most peaceful region, reached historically high levels of peace with fifteen of the twenty most peaceful countries (including Iceland, Denmark, and Austria). At the same time, increased civil unrest and terrorist activity rendered the Middle East and North Africa the world's least peaceful. (U.S. armed forces were active in the five least-peaceful countries—Syria, South Sudan, Iraq, Afghanistan, and Somalia.) The intensity of internal armed conflict increased dramatically. The economic impact of violence in 2015 reached $13.6 trillion—13.3 percent of global GDP, equal to the combined economies of Canada, France, Germany, Spain, and the United Kingdom.

The *Index* found that in countries with higher levels of "positive peace"—i.e., with attitudes, structures, and institutions that underpin peaceful societies—internal resistance movements were less likely to become violent and were more likely to successfully achieve concessions from the state.

The following list highlights some of the global initiatives designed to lead the world toward the goal of "positive peace."

A Green Geneva Convention

Numerous treaties have been created to protect portions of the global environment—the Kyoto Protocol, the International Framework Convention on Climate Change, the United Nations Law of the Sea, the Convention on

International Trade in Endangered Species of Wild Fauna and Flora—but the series of major laws governing war and humanitarianism, the Geneva Conventions, fall short of protecting the natural world from the flames of war. Dr. Klaus Toepfer (former executive secretary of the United Nations Environment Programme) was among the first to call for a Green Geneva Convention, arguing that environmental security must be a fundamental part of any enduring peace policy and that those who deliberately put the environment at risk in war should face trial and imprisonment.

www.unep.org

Green Constitutions

In 2008, Ecuador became the first country to incorporate the "rights of nature" in their constitution. Rather than treating nature as property, Ecuador believes that *Pachamama* (the Indigenous word for "Mother Earth") has a legal "right to exist." When environmental injuries occur, the ecosystem itself can be named as a defendant. Ecuador's "Green Constitution" contains the following protections:

Article 1. Nature or *Pachamama*, where life is reproduced and exists, has the right to exist, persist, maintain, and regenerate its vital cycles, structure, functions, and its processes in evolution. . . . The State will . . . promote respect towards all the elements that form an ecosystem.

Article 2. Nature has the right to an integral restoration.

Article 3. The State will apply precaution and restriction measures in all the activities that can lead to the extinction of species, the destruction of the ecosystems, or the permanent alteration of the natural cycles.

www.rightsofmotherearth.com/ecuador-rights-nature

U.S. Department of Peace

On September 14, 2005, then representative Dennis J. Kucinich introduced H.R. 3760, a bill to create a cabinet-level Department of Peace and Nonviolence to promote nonviolent conflict resolution as an organizing principle to help create conditions for a more peaceful world. The Department would advise the president on matters of national security, including the protection of human rights and the prevention and de-escalation of unarmed and armed international conflict.

http://peacealliance.org

The Nonviolent Peaceforce

The mission of the Belgium-based Nonviolent Peaceforce (NP) is to promote, develop, and implement unarmed civilian peacekeeping as a tool for reducing violence and protecting civilians in situations of violent conflict. NP peacekeeping teams—currently deployed in the Philippines, South Sudan, Myanmar, and the Middle East—include experienced peacekeepers, veterans of conflict zones, and trained volunteers. NP's Vision Statement reads, in part:

> Our activities have ranged from entering active conflict zones to remove civilians in the crossfire to providing opposing factions a safe space to negotiate. Other activities include serving as a communication link between warring factions, securing safe temporary housing for civilians displaced by war, providing violence prevention measures during elections, and negotiating the return of kidnapped family members.

www.nonviolentpeaceforce.org

Environmental Armies

Nicaragua has an Autonomy Law that recognizes Indigenous sovereignty and an Ecological Battalion in its army that serves to protect the Bosawas Rainforest, a UN-designated World Nature Preserve. The forest was under siege by colonizers, farmers, cattle-ranchers, loggers, and miners until the deployment of 580 "eco-soldiers" in 2012. The mission, Operation Green Gold, is Central America's first initiative to use soldiers to combat climate change by protecting Indigenous lands, wilderness, and native habitat. The armed eco-force has successfully reduced deforestation and illegal logging. Today 14 percent of Nicaragua's undeveloped land is under armed protection.

Military Accountability and Restoration

The Worldwatch Institute has called for the Pentagon to provide "all documentation pertaining to the environmental conditions of U.S. bases" at home and abroad and commit to long-term, post-closure cleanup agreements. This would include comprehensive environmental assessments, conducted in collaboration with democratically appointed representatives from host nations. Worldwatch also calls for contaminated nuclear sites and test ranges to be permanently sealed off as enduring reminders of the folly of nuclear proliferation.

www.worldwatch.org

Plowshares Initiatives

Plowshares policies would call for the diversion of tax dollars from weapons production to environmental restoration. If such policies were adopted, weapons labs would be required to cease military production and concentrate on mitigating the environmental damage caused by military activities. California's Lawrence Livermore National Laboratory (LLNL), a major designer of nuclear weapons, devotes just 2 percent of its budget to environmental restoration. In 1983, nearby residents formed Tri-Valley CAREs, a citizen's watchdog group. Tri-Valley CAREs monitors radioactive pollution released by the LLNL while campaigning to transform the site from a weapons plant to a research center that addresses real-world solutions to energy, food, and health problems.

www.trivalleycares.org

Peacebuilding Programs

In 2012, in response to growing pressure from peace and human rights activists, the U.S. State Department established an Atrocities Prevention Board. In 2015, the State Department declared: "Preventing mass atrocities is a core national security interest and moral responsibility of the U.S." The State Department now oversees a Bureau of Conflict and Stabilization Operations, a Complex Crisis Fund, and other "non-militarized strategies" that can be used in lieu of the armed escalations that still dominate U.S. foreign policy. As the Friends Committee on National Legislation notes, "Investing early to prevent war is far more cost-effective than military intervention after a crisis erupts." Proactive peacebuilding efforts have proven successful in Kenya, Burundi, Sri Lanka, and Guinea, but continued funding for these programs remains uncertain.

www.fcnl.org

The Peace Dividend

With the bulk of the U.S. budget supporting costly military programs (while cuts are made in food stamps, social security, housing, and education), it's time to consider a question posed by UNESCO: "What does the world want and how can we pay for it using military expenditures?" In 1978, UNESCO calculated that redirecting 30 percent of the world's military expenditures ($780 billion) could solve most of the planet's persistent problems, including: eliminating starvation; providing health care; offering shelter to the homeless; guaranteeing clean safe water; eliminating illiteracy; providing clean, safe, renewable energy; retiring all developing nation debts; stabilizing population growth; stopping erosion; halting deforestation; preventing ozone loss and acid rain; addressing climate change; removing land mines; providing refugee relief; eliminating nuclear weapons; and building democracy.

www.unesco.org; www.worldgame.org

National Budgets for People, Not Plunder

In 2015, the International Institute for Strategic Studies reported the Pentagon's $581 billion budget accounted for more than one-third of *all* military spending on Earth. For every $1 billion spent on the military, twice as many jobs could be created in the civilian sector. Investing in infrastructure—roads and bridges, schools, hospitals, clean water, and renewable energy—is wiser than pouring the money into single-use items like bullets and bombs. One billion federal dollars invested in the military creates 11,200 jobs: the same investment in clean energy could yield 16,800 jobs. The Borgen Project estimates that ending world hunger would cost $30 billion per year—the same amount the Pentagon burns through in eight days.

The Global Campaign on Military Spending is calling for the transfer of military money to fund five broad areas critical to a global transformation toward a culture of peace: disarmament and conflict prevention; sustainable development and anti-poverty programs; climate stabilization and biodiversity loss; public services/social justice, human rights, gender equality, and green job–creation; and humanitarian programs for the disadvantaged.

http://demilitarize.org

Nuclear Disarmament

Since 1999, Congresswoman Eleanor Holmes Norton (D-DC) has repeatedly introduced a Nuclear Disarmament and Economic Conversion Act (NDECA) requiring the United States to "dismantle its nuclear weapons" and redirect the savings "to address human and infrastructure needs such as housing, health care, education, agriculture, and the environment." The NDECA would require the United States to "undertake vigorous, good-faith efforts to eliminate war, armed conflict, and all military operations." Norton's campaign has won the support of Physicians for Social Responsibility and the International Physicians for the Prevention of Nuclear War (winner of the 1998 Noble Peace Prize).

www.psr.org; www.ippne.org

The Abolition of Weapons of Mass Destruction

Despite many calls for nuclear disarmament, the United States has done little to reduce or rein in the nuclear threat. There are treaties banning biological and chemical weapons, but there are no treaties banning nuclear weapons. Nuclear arms control agreements have reduced the world's atomic arsenal from 56,000 to 16,300, but the White House (under both Obama and Trump) has called for spending $1 trillion on a new generation of atomic weapons and delivery systems. The International Campaign to Abolish Nuclear Weapons (ICAN) is working toward multilateral negotiations for a treaty banning nuclear weapons by engaging with humanitarian, environmental, human rights, peace, and development organizations in more than ninety countries.

www.icanw.org

The Humanitarian Pledge

The Humanitarian Pledge was issued on December 9, 2014, at the conclusion of the Vienna Conference on the Humanitarian Impact of Nuclear Weapons, attended by 158 nations. This important document provides governments with the opportunity to move beyond fact-based discussions on the effects of nuclear weapons and into the start of treaty negotiations. As of February 2017, 127 nations had endorsed the Pledge.

www.icanw.org/pledge

Nuclear-Free Nations and Nuclear-Weapons-Free Zones

In 1987, thanks to the influence of Dr. Helen Caldicott, the first president of Physicians for Social Responsibility, New Zealand's Labour government passed a Nuclear Free Zone, Disarmament and Arms Control Act—thereby making New Zealand the world's first nuclear-free nation. On February 28, 2000, Mongolia became the world's second self-declared nuclear-weapon-free nation. There are five treaties establishing nuclear-weapons-free zones (NWFZ): the Treaty of Tlatelolco (Latin America and the Caribbean); the Treaty of Rarotonga (South Pacific): the Treaty of Bangkok (Southeast Asia); the Treaty of Pelindaba (Africa); and the Treaty on a Nuclear-Weapon-Free Zone in Central Asia. These NFWZs now cover 57 million square miles—56 percent of the planet's land area. There also is a Treaty on the Prohibition of the Emplacement of Nuclear Weapons and Other Weapons of Mass Destruction on the Sea-Bed and the Ocean Floor.

Establish Zones of Peace

The concept of recognizing special conflict-free sanctuaries goes back to the ancient Egyptians, Polynesians, and Hebrews and remains prevalent today. In 1987, the United Nations designated a vast stretch of the South Atlantic—reaching from the coast of Africa to the shores of South America—a "Zone of Peace." In January 2014, thirty-three Latin American and Caribbean nations met in Cuba to renounce the use of war and proclaim a regional Zone of Peace. In December 2014, India called for designating the Indian Ocean a Zone of Peace. And there is an ongoing campaign to declare the Middle East a nuclear-free zone.

www.uia.org/s/or/en/1100016029

Abolish the Arms Trade

The $70 billion-per-year trade in weapons contributes to the escalation of violence and terrorism. The global arms trade aids dictatorships, creates international instability, and perpetuates the belief that peace can be achieved by arms. The world's main arms exporters are the United States, Russia, Germany, France, and the United Kingdom. Arms manufacturers enjoy federal subsidies and lucrative government contracts, and they are free to sell weapons on the open market—sometimes arming both sides of a conflict or even "enemy" forces (as when the U.S.-armed Mujahedeen evolved into al-Qaeda, and U.S. arms for Iraq ended up in the hands of ISIS). As part of the UN mandate to protect "international peace and security," the Campaign Against Arms Trade and other peace groups are calling on UN Security Council to add the arms trade to the International Criminal Court's list of "crimes against humanity."

www.caat.org.uk; www.amnesty.org/en/what-we-do/arms-control

Enforce an International Arms Code of Conduct

In 1999, a commission of Nobel Peace Prize winners led by former Costa Rican president Óscar Arias drafted legislation to control global arms sales. The Arms Code of Conduct requires that countries wishing to buy arms must first meet certain criteria—including a respect for democracy and human rights. The code, which bans sales to dictatorships and oppressive regimes, poses a dilemma for America's "military-industrial complex." According to Demilitarization for Democracy [www.dfd.net], the Clinton Administration exported $8.3 billion in arms to fifty-two "non-democratic regimes"—including Saudi Arabia, Kuwait, Egypt, Thailand,

and Pakistan—while providing military training to forty-seven dictatorships. The Obama Administration provided billions of dollars of weapons to oppressive regimes in Bahrain, Egypt, Iraq, Jordan, Kuwait, Saudi Arabia, Tunisia, and the United Arab Emirates. Members of the European Parliament have challenged the United States to abide by an Arms Trade Code of Conduct adopted by the European Union.

Replace War Zones with Peace Parks

The planet's first peace park, the Waterton-Glacier International Peace Park, was established in 1932 to mark the friendship between the United States and Canada. There are now 143 peace parks in 42 countries. The United Nations and the International Union on the Conservation of Nature require that peace parks promote nonviolence and biodiversity in areas that have experienced "significant conflict." After a 1995 border war between Peru and Ecuador in the Cordillera del Condor (a mountainous region rich in biodiversity), environmentalists and indigenous groups brokered a peace treaty that created parks on both sides of the disputed border and set the stage for a shared binational park that has served to ensure a lasting peace. Today, eighteen peace parks straddle countries with shared borders. Large transboundary peace parks link Botswana, South Africa, Zimbabwe, and Mozambique. Other parks share the borders of Albania/Kosovo/Montenegro, China/Pakistan, Costa Rica/Panama, and Mexico/United States. West Africa's "W" International Peace Park has helped to reduce wildlife poaching and deforestation in the Niger River Basin. The Selous-Niassa Elephant Corridor protects critical wildlife areas between Tanzania and Mozambique.

www.peaceparks.org

Create a Global Security System: An Alternative to War

Over the last 150 years, revolutionary new methods of nonviolent conflict management have been developed to end warfare. The Alternative Global Security System would replace the failed weapons-based national security approach with a concept of "common security" based on three broad strategies: (1) demilitarizing security, (2) managing conflicts without violence, and (3) creating a culture of peace. World Beyond War's 2015 book, *A Global Security System: An Alternative to War*, describes the "hardware" of creating a peace system and the "software" (values and concepts) needed to *operate* and expand the system globally. The program includes detailed strategies to demilitarize countries, manage conflicts peacefully, reduce poverty, and promote environmental stewardship.

http://worldbeyondwar.org/alternative/

Establish Institutions of Nonviolence

In *Why Civil Resistance Works* (2011), Erica Chenoweth and Maria Stephan demonstrated that over a span from 1900 to 2006, nonviolent resistance was twice as successful as armed resistance in creating stable democracies, with less chance those states revert to violence. History offers numerous examples of unpopular governments removed through nonviolent struggle—from Mahatma Gandhi's campaign against the British to the overthrow of the Marcos regime in the Philippines and the rise of the Arab Spring. In 2015, the Nobel Peace Prize was awarded to the Tunisian National Dialogue Quartet, a coalition made up of a labor union, a trade confederation, a human rights league, and a lawyers group that used nonviolence to establish a "peaceful political process at a time when the country was on the brink of civil war."

www.aforcemorepowerful.org

Demilitarize National Security

The Pentagon's global footprint of foreign bases, naval fleets, missile installations, and military interventions feeds hostility abroad. "Defense spending" primarily directed at "projecting U.S. military power worldwide" is not defensive. A first step toward demilitarizing national security would be to establish a "non-provocative defense" limited to defending national borders. A defensive military posture would eliminate long-range bombers, nuclear submarines, carrier fleets, and intercontinental missiles. Twenty-two countries have disbanded their militaries. Costa Rican president José Figueres Ferrer abolished the military in 1948 and invested heavily in cultural preservation, environmental protection, and public education. Costa Rica now boasts one of the highest literacy rates in the Americas.

www.peacesystems.org; peacemagazine.org

Close Foreign Bases

The United States has between 700 and 1,180 bases in more than 60 countries around the world. Eliminating foreign military bases—a goal of the Alternative Global Security System—goes hand in hand with non-provocative defense. More than seventy years after World War II, the

United States continues to maintain twenty-one military bases in Germany and twenty-three bases in Japan—bases that are a frequent source of resentment among local residents. Withdrawing from the military occupation of foreign countries would save billions and reduce the War System's ability to inflame global insecurity.

www.tni.org/en/publication/foreign-military-bases-and-the-global-campaign-to-close-them;

www.no-bases.org

Halt the Use of Militarized Drones

U.S. drones are regularly used for targeted killings in Pakistan, Yemen, Afghanistan, and Somalia. The justification for these attacks (which have killed hundreds of innocent civilians) is the questionable doctrine of "anticipatory defense." The White House claims the authority to order the death of anyone deemed a terrorist threat—even if the United States has not declared war on the countries attacked; even if the targets are U.S. citizens. Targeted killings violate the Constitution's guarantee of due process. Ultimately, drones are counterproductive. As Gen. Stanley McChrystal (former commander of U.S. and NATO Forces in Afghanistan) has observed: "For every innocent person you kill, you create ten new enemies."

www.stopdrones.com; http://nodronesnetwork.blogspot.com

End Invasions and Occupations

The occupation of one country by another is a fundamental threat to security and peace that can provoke resentment and local resistance, ranging from street protests to armed resistance and "terrorist" assaults. The resulting conflicts often kill more civilians than insurgents while creating floods of refugees. The UN Charter outlaws invasion (unless they are in retaliation for a prior invasion—an inadequate provision). The presence of troops of one country inside another—with or without an invitation—destabilizes global security and makes armed conflicts more likely. Invasions and occupations would be prohibited under an Alternative Global Security System.

www.un.org/en/charter-united-nations/index.html

Dismantle Military Alliances

The North Atlantic Treaty Organization is a Cold War relic. The Warsaw Pact was disbanded following the collapse of the Soviet Union, but NATO (in violation of Western promises) has continued to expand and now encroaches on Russia's borders. NATO has undertaken military exercises well beyond Europe's borders—in Eastern Europe, North Africa, and the Middle East. An Alternative Global Security System would replace NATO with new international institutions designed to manage conflict without violence.

http://worldbeyondwar.org/outline-alternative-security-system

Promote Fair and Sustainable Global Economies

Social injustice, high youth unemployment, and economic desperation can create a seedbed for extremists. The imbalance between the affluence of the Global North and the poverty of the Global South could be righted by democratizing international economic institutions and taking care to conserve the ecosystems upon which all economies rest. Competition for limited resources stokes tensions between nations and *within* nations. Using natural resources more efficiently and developing non-polluting technologies can reduce ecological stress. As the former UN under secretary general for Disarmament Affairs Jayantha Dhanapala has said: "Wars claimed more than 5 million lives in the 1990s, and nearly 3 billion people, almost half the world's population, live on a daily income of less than $2 a day. Poverty and conflict are not unrelated; they often reinforce each other. . . . Even where there is no active conflict, military spending absorbs resources that could be used to attack poverty."

www.un.org/disarmament/education/activities.html

Promote a Global, Green Marshall Plan

A Global Marshall Plan (GMP) designed to achieve economic and environmental justice worldwide could democratize international economic institutions. The goal would be similar to the UN Millennium Development Goals: to end poverty and hunger, develop local food security, provide education and health care, and achieve stable, efficient, sustainable economic development. To prevent the GMP from becoming a policy tool of rich nations, the work could be administered by an independent, international nongovernmental organization. The GMP would require strict accounting and transparency from recipient governments.

www.globalmarshallplan.org

Replace the United Nations with an Earth Federation

The United Nations' failures to solve the planetary threats facing humankind are due to its very nature. The United Nations has no legislative powers—it cannot enact binding laws. It has failed to solve the problems of social and economic development, and global poverty remains acute. It has not stopped deforestation, climate change, fossil fuel use, global soil erosion, or ocean pollution. Instead of promoting disarmament, the United Nations requires members to maintain armed forces that can be called upon for "peacekeeping" missions. The World Court has no power to bring disputes before it. The General Assembly can only "study and recommend" and lacks the power to change anything. The United Nations must either evolve or be replaced, perhaps by a nonmilitary Earth Federation composed of a democratically elected World Parliament with power to pass binding legislation.

www.earthfederation.info; http://worldparliament-gov.org

Create Cultures of Peace

In 1999, the United Nations General Assembly approved a Programme of Action on a Culture of Peace. Article I called for: "Respect for life, ending of violence and promotion and practice of non-violence through education, dialogue and cooperation; . . . [c]ommitment to peaceful settlement of conflicts." The UN General Assembly has identified eight action areas: fostering a culture of peace through education: promoting sustainable economic and social development; promoting respect for all human rights; ensuring equality between women and men; fostering democratic participation; advancing understanding, tolerance, and solidarity; supporting participatory communication and the free flow of information and knowledge; and promoting international peace and security. The Global Movement for the Culture of Peace, founded by UNESCO in 1992, is a partnership of civil society groups working to make militarism obsolete.

www.culture-of-peace.info/copoj/

Encourage Peace Journalism

The prowar bias commonly seen in schoolroom history texts also infects mainstream journalism where many reporters, columnists, and news pundits promote the fable that war is inevitable and that it brings peace. "Peace journalism" (conceived by scholar Johan Galtung) encourages editors and writers to explore nonviolent alternatives to conflict. In contrast to war journalism's "good guys versus bad guys" approach, peace journalism focuses on the structural and cultural causes of violence and explores peace initiatives commonly ignored by the mainstream media.

Examples include the Center for Global Peace Journalism's *Peace Journalist* magazine and the Oregon Peace Institute's *PeaceVoice.*

www.park.edu/center-for-peace-journalism; http://blog.orpeace.us

Practice Nonviolent War-Tax Resistance

> *Let them march all they want, as long as they continue to pay their taxes.*
>
> —Alexander Haig, U.S. Secretary of State, June 12, 1982

War-tax refusal has a long tradition among religious and secular opponents of war—including Quakers, Mennonites, and Brethren. Conscientious Americans have refused to pay taxes for virtually every U.S. war. (Henry David Thoreau was famously jailed for refusing to finance Washington's war on Mexico.)

In 1984, the War Resisters League issued an Appeal to Conscience that read, in part:

> It is clear that the U.S. government's ability to threaten, coerce, and, if deemed necessary, make war on other nations is a direct result, not only of our economic might, but also the unprecedented size of our military arsenal, which is now far larger than that of all our allies and "enemies" combined. It is equally clear that the maintenance of this arsenal depends upon the willingness of the American people—through their federal tax payments—to finance it. . . . Refusal to pay taxes used to finance unjust wars, along with refusal by soldiers to fight in them, is a direct and potentially effective form of citizen noncooperation, and one that governments cannot ignore.

www.warresisters.org

Encourage Planetary Citizenship: One People, One Planet, One Peace

Homo sapiens constitute a single species, with a marvelous diversity of ethnic, religious, economic, and political systems. With climate-stressed global disasters already under way—including massive deforestation and unprecedented rates of extinction—we face a planetary emergency. We need to place the long-term health of the global commons above the short-term goals of national interest. Protecting the commons is best achieved by voluntary consensus and a recognition of mutual respect that arises out of a sense of responsibility for the planet's well-being. Conflict does not have to lead to war. We have established nonviolent methods of conflict resolution that can provide for common security—a world free from fear, want, and persecution, and a civilization in balance with a healthy biosphere.

http://planetarycitizens.net

INTERNATIONAL PEACE AND ENVIRONMENTAL ORGANIZATIONS

Albert Einstein Institution, www.aeinstein.org

American Friends Service Committee, www.afsc.org

Amnesty International, www.amnesty.org/en/

Campaign Against the Arms Trade, www.caat.org.uk/

Campaign to Close All Military Bases, www.tni.org/es/node/2759

Campaign Nonviolence, www.paceebene.org

Carnegie Endowment for International Peace, www.carnegieendowment.org

Carter Center, www.cartercenter.org/peace

Christian Peacemaker Teams, www.cpt.org

Citizens for Global Solutions, www.globalsolutions.org

Coalition Against the Arms Trade, www.caat.org.uk

Coalition for Peace Action, www.peacecoalition.org

EarthAction, www.earthaction.org

Environmentalists Against War, www.envirosagainstwar.org

Fellowship of Reconciliation, www.forusa.org

Greenpeace International, www.greenpeace.org

Hague Appeal for Peace, www.haguepeace.org

Human Rights Watch, www.hrw.org

Institute for Inclusive Security, www.inclusivesecurity.org

International Campaign to Abolish Nuclear Weapons, www.icanw.org

International Center for Nonviolent Conflict, www.nonviolent-conflict.org

International Committee of the Red Cross, https://www.icrc.org

International Criminal Court, www.icc-cpi.int

International Fellowship of Reconciliation, www.ifor.org

International Peace Research Association, www.iprapeace.org

International Peace Bureau, www.ipb.org

International Physicians for the Prevention of Nuclear War, www.ippnw.org

International Union for the Conservation of Nature, www.iucn.org/

Jewish Peace Fellowship, www.jewishpeacefellowship.org

Journal of Peace Education, www.tandfonline.com/loi/cjpe20

Mayors for Peace, www.mayorsforpeace.org

Muslim Peace Coalition USA, http://muslimpeacecoalition.org

National Network Opposing the Militarization of Youth, nnomy.org

National Peace Foundation, www.nationalpeace.org

Nobel Women's Initiative, www.nobelwomensinitiative.org

Nonviolence International, www.nonviolenceinternational.net

Nonviolent Peaceforce, www.nonviolentpeaceforce.org

Nukewatch, www.nukewatchinfo.org

Oxfam International, www.oxfam.org

Pace e Bene, www.paceebene.org

Pax Christi International, www.paxchristi.net

Peace Action, www.peace-action.org

Peace Brigades International, www.peacebrigades.org

Peace and Justice Studies Association, www.peacejusticestudies.org

Peace Journalism, www.peacevoice.info; www.park.edu/center-for-peace-journalism

Peace People, www.peacepeople.com

Physicians for Social Responsibility, www.psr.org

Ploughshares Fund, www.ploughshares.org

Stockholm International Peace Research Institute, www.sipri.org

Transcend International, www.transcend.org

United for Peace and Justice, www.unitedforpeace.org

United Nations Association of the United States, www.unausa.org

United Nations Environment Programme, www.unep.org

Veterans for Peace, www.veteransforpeace.org

Waging Nonviolence, www.wagingnonviolence.org

War Resisters League, www.warresisters.org

War Resisters International, www.wri-irg.org

Women's Action for New Directions, www.wand.org

Women's International League for Peace and Freedom, www.wilpfinternational.org

World Beyond War, worldbeyondwar.org

World Federalist Movement, wfm-igp.org

Worldwatch Institute, www.worldwatch.org

World Parliament, www.worldparliament-gov.org

World Peace Council, www.wpc-in.org

ABOUT THE CONTRIBUTORS

WILLIAM J. ASTORE is a professor of history at the Pennsylvania College of Technology and a retired lieutenant colonel (United States Air Force). He served as the director of International History at the U.S. Air Force Academy and is the author of three books, including *Soldiers' Lives through History: The Early Modern World* (Greenwood Press). He can be reached at wastore@pct.edu.

MEDEA BENJAMIN is founding director of Global Exchange and cofounder of Code Pink. Benjamin has been arrested numerous times in nonviolent antiwar protests in the United States and around the world. She has been honored with the Martin Luther King Jr. Peace Prize, the Marjorie Kellogg National Peacemaker Award, and the Gandhi Peace Award for her "unyielding advocacy for social justice of more than thirty years." She is the author of numerous books, including *Cuba: Talking about Revolution*, *The Greening of the Revolution*, and *No Free Lunch*. Her latest book is *Drone Warfare: Killing by Remote Control.*

LISA BRADY is associate professor of history at Boise State University. She specializes in North American and global environmental history and is the author of *War upon the Land: Military Strategy and the Transformation of Southern Landscapes during the American Civil War.* Brady has written extensively about the history of the Korean DMZ and other issues involving military destruction in the modern age.

PIERCE BROSNAN is an Irish-American actor, film producer, artist, philanthropist, and two-time Golden Globe Award nominee. He also is an ardent environmental activist who has worked closely with the Natural Resources Defense Council to defend marine mammals from the harm caused by military naval activities. Brosnan has been honored with the Robert Graham Visionary Award and the Golden Kamera Award for his environmental work. He has also served as a UNICEF Ireland Ambassador since 2001.

DAVID BROWER was a legendary mountaineer (with more than seventy first ascents to his credit), a combat veteran of the 10th Mountain Division in World War II, the first executive director of the Sierra Club, and founder of the League of Conservation Voters, Friends of the Earth, and the Earth Island Institute. At the Sierra Club, Brower introduced a series of photo-illustrated large-format books that became powerful campaign tools in campaigns to protect the wild. He authored a number of books, including *For Earth's Sake: The Life and Times of David Brower* and *Let the Mountains Talk, Let the Rivers Run* (the first commercial book in the United States printed on tree-free kenaf plant paper).

MAJOR-GENERAL SMEDLEY DARLINGTON BUTLER was the most highly decorated soldier in Marine Corps history. Awarded two Congressional Medals of Honor for heroism, he ultimately renounced war in his broadside, "War Is a Racket." In 1934, Butler became a Congressional whistleblower when he exposed a secret plan (hatched by the American Legion and the heads of DuPont, Goodyear, Bethlehem Steel, and J.P. Morgan) to seize the White House and overthrow President Franklin Roosevelt. The plot was prompted by anger over Roosevelt's New Deal social programs. The McCormack-Dickstein Committee confirmed the plot but suppressed the details about who was behind it.

DR. HELEN CALDICOTT has devoted more than half her life to educating the public about the medical hazards of the nuclear age. In 1980, she left the staff of Boston's Children's Hospital Medical Center to found the Nuclear Policy Research Institute (now known as Beyond Nuclear). She served as president of Physicians for Social Responsibility for nearly a decade before going on to work with the International Physicians for the Prevention of Nuclear War. The Smithsonian named Dr. Caldicott "one of the most influential women of the twentieth century." She has authored seven books, including *Nuclear Madness* and *If You Love This Planet: A Plan to Heal the Earth*. Her latest book is *Crisis without End*.

MARJORIE COHN is a professor at Thomas Jefferson School of Law, past president of the National Lawyers Guild, and deputy secretary general of the International Association of Democratic Lawyers. Her most recent book is *Drones and Targeted Killing: Legal, Moral, and Geopolitical Issues*. More information is available at www.marjoriecohn.com.

JAMES CORBETT is the host of *The Corbett Report*, an independent, listener-supported alternative news source. Based in Japan, he hosts a weekly podcast and several online video series. He is an editorial writer for *The International Forecaster*, a weekly e-newsletter. His YouTube videos have garnered more than 22 million views.

SALLY JEWELL COXE is president and cofounder of the nonprofit Bonobo Conservation Initiative, based in Washington, DC, and Kinshasa, Democratic Republic of the Congo. BCI and its partners have launched a global campaign to save bonobos by establishing protected areas, supporting community conservation, and promoting female empowerment. BCI works with the DRC and the United Nations to protect the bonobos, the biodiversity of the Congo Forest, the local people, and their prospects for peace. More information is available at www.bonobo.org.

DENIS DELESTRAC is an award-winning film director and writer whose documentaries have provoked public debate and influenced political decision-making internationally. After launching his career as a writer and photographer in the United States, Delestrac stepped by chance into filmmaking in 2001. In 2009, he teamed with Mark Achbar (director of *The Corporation*) to produce *Pax Americana and the Weaponization of Space*. His latest film, *Sand Wars*, has won eleven awards including a Gold Panda, the Greenpeace Prize, and a Gemini Award.

KATIE DRUMMOND is the science editor at *The Verge*, Vox Media's online journal of technology and culture. Drummond previously covered military technology for *Wired* magazine's Danger Room. She has covered breaking science and health news for *The New Republic*, *New York* magazine, and *Popular Science*.

DANIEL ELLSBERG has worked in the State and Defense departments under Presidents Kennedy, Johnson, and Nixon. In 1971, he released the *Pentagon Papers* to the press. Ellsberg is a senior fellow at the Nuclear Age Peace Foundation and is currently a Foundation Distinguished Fellow. He is the author of *Secrets: A Memoir of Vietnam and the Pentagon Papers* and the forthcoming book, *The Doomsday Machine: Confessions of a Nuclear War Planner* (Bloomsbury, 2018).

ROBERT FISK, a British journalist and best-selling author, has covered wars in the Middle East for *The Times of London* and *The Independent* since 1976. Known for his passionate journalism and his readiness to take risks to report a story, Fisk was one of the few Western reporters to interview al-Qaeda leader Osama bin Laden. Fisk has been honored by the British Press Awards as International Journalist of the Year and Foreign Reporter of the Year. He is the author of several books, including *The Point of No Return, In Time of War, Pity the Nation: Lebanon at War*, and *The Great War for Civilization: The Conquest of the Middle East.*

JOSHUA FRANK is managing editor of *CounterPunch*, author of *Left Out! How Liberals Helped Reelect George W. Bush*, and, along with Jeffrey St. Clair, the coeditor of *Red State Rebels: Tales of Grassroots Resistance in the Heartland*, and *Hopeless: Barack Obama and the Politics of Illusion*, published by AK Press.

PHILLIP FRAZER published and wrote the environmental newsletter *News on Earth*. He has published, edited, and written for other US publications including *The Hightower Lowdown* and, in Australia, *The Digger* and the Australian edition of *Rolling Stone*. He currently lives in Coorabell, Australia.

DAVE GILSON is a senior editor at *Mother Jones* magazine, where he has worked since 2003. Read more of his writing and reporting at www.davegilson.com.

Dr. JANE GOODALL, DBE, is an ethologist, famous for her research into wild chimpanzees in Tanzania—research that continues to this day and now spans nearly sixty years. She is founder of the Jane Goodall Institute and a UN Messenger of Peace. Roots & Shoots is the Jane Goodall Institute's global program for young people of all ages and empowers them to become involved in hands-on projects for their local community, for animals and for the environment we all share. More information is available at www.janegoodall.org and www.rootsandshoots.org.

KARL GROSSMAN is a full professor of journalism at the State University of New York/College at Old Westbury. He the recipient of the George Polk and John Peter Zenger reporting awards and the author of *The Wrong Stuff: The Space Program's Nuclear Threat to Our Planet.* He is the host of the nationally aired TV program *Enviro Close-Up*, which has produced TV documentaries, including *Three Mile Island Revisited, The Push to Revive Nuclear Power*, and *Nukes in Space: The Nuclearization and Weaponization of the Heavens.*

HUGH GUSTERSON is a professor of anthropology and sociology at George Mason University. His expertise is in nuclear culture, international security, and the anthropology of science. He is the author of *Nuclear Rites: A Weapons Laboratory at the End of the Cold War* (University of California Press, 1996) and *People of the Bomb: Portraits of America's*

Nuclear Complex (University of Minnesota Press, 2004). He also coedited *Why America's Top Pundits Are Wrong* and *The Insecure American.*

TOM H. HASTINGS is director of PeaceVoice (a program of the Oregon Peace Institute), director of peace and nonviolence studies at Portland State University, and a member of the core faculty at PSU's Conflict Resolution graduate program. He has served three prison terms for nonviolently protesting war and social injustice. His books include *Lessons of Nonviolence, Nonviolent Response to Terrorism,* and *Ecology of War and Peace: Counting Costs of Conflict.*

HELEN JACCARD is cochair of the Veterans for Peace Nuclear Abolition Working Group and chair of the organization's Environmental Cost of War and Militarism Working Group. She has traveled widely to conduct research on the environmental and health effects of the military activities. She is also a researcher for the campaign to clean up abandoned uranium mines in the United States.

ANN JONES is a journalist, photographer, and author of ten books, including *They Were Soldiers: How the Wounded Return from America's Wars—the Untold Story* and two books about the impact of war on civilians: *Kabul in Winter* and *War Is Not Over When It's Over.* She has reported from Afghanistan, Africa, Southeast Asia, and the Middle East and embedded with U.S. forces in Afghanistan to report on the damage done to America's soldiers. She is now an associate of the Charles Warren Center for Studies in American History at Harvard University. Her website is www.annjonesonline.com.

DARLENE KEJU-JOHNSON was born on the island of Ebeye in the Marshall Islands in 1951, downwind from the Bikini and Enewetak atolls where the United States tested sixty-seven nuclear weapons. Her book, *Don't Ever Whisper* (coauthored with her husband Giff Johnson, editor of the *Marshall Islands Journal*), was the first to bring the suffering of the Marshall Islanders to the attention of the world. A powerful and tireless speaker for Pacific Islanders harmed by nuclear testing, she died in 1996, at the age of forty-five, from breast cancer.

ELIZABETH KEMF is a Swiss-American journalist who first traveled to Vietnam in 1985, at a time when the country was still closed to most foreign journalists. She has returned to Vietnam numerous times to report on postwar restoration efforts. Dr. Kemf's professional activities include environmental and anthropological research, writing, editing, and filmmaking. She is the author of two nonfiction books, including *Month of Pure Light: The Regreening of Vietnam*. Kemf lives in Switzerland, where she works for the Worldwide Fund for Nature.

MICHAEL T. KLARE is a Five Colleges professor of Peace and World Security Studies, teaching at Hampshire, Amherst, Smith, Mount Holyoke, and the University of Massachusetts, Amherst. Klare is a defense correspondent for *The Nation* and serves on the boards of directors of Human Rights Watch and the Arms Control Association. He is the author of *Resource Wars, The Race for What's Left* and *Blood and Oil: The Dangers and Consequences of America's Growing Petroleum Dependency.*

WINONA LADUKE is an enrolled member of the Mississippi Band Anishinaabeg who lives and works on the White Earth Reservations. She is an internationally renowned Native American activist and advocate for environmental, women's, and children's rights. A graduate of Harvard and Antioch Universities, with advanced degrees in rural economic development, LaDuke is founder and executive director of Honor the Earth, a national advocacy group encouraging public support and funding for native environmental initiatives.

SUSAN D. LANIER-GRAHAM graduated magna cum laude from University of Maryland, with a degree in government and politics/comparative economics, and she holds a master of liberal arts with honors from Regis University in Denver, Colorado. An award-winning travel writer for more than two decades, Lanier-Graham is the host of WanderWithWonder.com. She is based in Arizona and has authored more than seventy-five books, including *The Ecology of War: Environmental Impacts of Weaponry and Warfare.*

JAMES LERAGER is the director of the Documentary Photography and Research Project and photography editor for *Latin American Perspectives.* Lerager holds a master's degree from the Goldman School of Public Policy from University of California at Berkeley and was a Fulbright Senior Scholar in Medellín, Colombia. He has presented thirty-five solo photography exhibitions at universities, museums, and galleries in the United States, South America, and Europe. He is the author of *In the Shadow of The Cloud, Nuclear History: Nuclear Destiny*, and *Mexico: Portraits of Complexity.*

JON LETMAN is an independent freelance journalist and photographer on the Hawaiian island of Kaua'i. His articles on conservation, the environment, politics, and the Asia-Pacific region have been published in *Al Jazeera, Truthout, Inter Press Service*, and the *Christian Science Monitor* and have appeared in publications in Finland, Iceland, Russia, Japan, Canada, the United Kingdom, and across the United States.

PETRA LOESCH is a Senior Clinical Science Liaison and Senior Clinical Research Program Manager at Massachusetts General Hospital's cancer center. She was formerly the Associate Director Medical Communications at Merrimack Pharmaceuticals. A native of Freiburg, Germany (a town that was mistakenly bombed by German pilots during World War II), Loesch first reported on the environmental impacts of low-level military jet flight when she was on the staff of *Earth Island Journal*.

CATHERINE LUTZ is the Thomas J. Watson Jr. Family Professor in Anthropology and International Studies at the Watson Institute for International Studies at Brown University. She is the author of *The Bases of Empire: The Global Struggle against US Military Posts* and *Bomb after Bomb: A Violent Cartography* (with Elin O'Hara Slavick, Carol Mavor, and Howard Zinn).

JERRY MANDER is a media activist and the author of *Four Arguments for the Elimination of Television* and *In the Absence of the Sacred*. As a partner and president of Freeman, Mander and Gossage, Mander worked with Sierra Club president David Brower to mount ad campaigns that blocked construction of dams in the Grand Canyon. In 1971, Mander founded Public Interest Communications, the first nonprofit advertising agency in the United States. He served as founder and executive director of the International Forum on Globalization for fifteen years. His most recent book is *The Capitalism Papers*.

MARGARET MEAD was an anthropologist who authored more than twenty books, including *Coming of Age in Samoa, Growing Up in New Guinea*, and *New Lives for Old*. One of the first to examine human development from a cross-cultural perspective, Mead challenged the prevailing belief that "primitive" people were childlike and inferior and demonstrated that violent conflict is not endemic to all cultures. Mead served as president of the American Anthropological Association and the American Association for the Advancement of Sciences and was awarded the Presidential Medal of Freedom.

KOOHAN PAIK is the coordinator for the Asia-Pacific Program at the International Forum on Globalization. She is also an activist, a writer, and a filmmaker. She is coauthor with Jerry Mander of *The Superferry Chronicles: Hawaii's Uprising against Militarism, Commercialism and the Desecration of the Earth.*

BARRY SANDERS is a Senior Fulbright Scholar and the author of *The Green Zone*, which Project Censored called "one of the Top Ten Censored Stories of 2009." He has twice been nominated for the Pulitzer Prize and is the author of fourteen books. He was the first to occupy the Gold Chair at Pitzer College and is founding cochair of the Master of Arts in Critical Theory and Creative Research at the Hallie Ford School of Graduate Studies, Pacific Northwest College of Art.

TYRONE SAVAGE is a Fulbright Scholar from Cape Town, South Africa, with a dual master's degree in international relations and public administration from Syracuse University. He has lectured at Cape Town's Stellenbosch University and has conducted research with the International Center for Transitional Justice, the Institut Français d'Afrique du Sud, the Institute for Justice and Reconciliation, and the United Nations Assistance Mission for Iraq. Savage has served as head of research and analysis in the Transitional Justice in Africa Program at the Institute for Justice and Reconciliation in Cape Town. He has published widely on the process of conflict transformation.

GAR SMITH, a veteran of Berkeley's Free Speech Movement, founded one of the first environmental organizations on a U.S. university campus. He worked with legendary environmentalist Dave Brower at Friends of the Earth and at Earth Island Institute and was the founding editor of *Earth Island Journal*. During a long career as an investigative reporter, Smith has won several Project Censored Awards. He currently serves as the director of Environmentalists Against War and is the author of *Nuclear Roulette*.

SUSI SNYDER is the Nuclear Disarmament Program Manager for PAX, a Netherlands-based peace organization. She serves on the International Steering Group of the International Campaign to Abolish Nuclear Weapons and previously served as the secretary general of the Women's International League for Peace and Freedom. WILBERT VAN DER ZEIJDEN is the senior researcher of the Security and Disarmament team of PAX. More information is available at www.paxforpeace.nl.

VANDANA SHIVA is a physicist, a philosopher, a feminist, and a world-renowned environmental leader. Director of the Research Foundation on Science, Technology, and Ecology, she is the author of many books, including *Stolen Harvest: The Hijacking of the Global Food Supply,* and *Soil Not Oil: Environmental Justice in an Age of Climate Crisis.* She is the founder of Navdanya, a grassroots movement promoting biodiversity and the use of native seeds, and is a recipient of the prestigious Right Livelihood Award. She holds a master's degree in the philosophy of science and a PhD in particle physics.

DAVID SWANSON is an author, activist, journalist, and radio host. He is the executive director of World Beyond War and campaign coordinator for RootsAction.org. Swanson's many books include *War Is a Lie, The Military Industrial Complex at 50, War No More: The Case for Abolition,* and *Killing Is Not a Way of Life.* He hosts *Talk Nation Radio* and blogs at DavidSwanson.org and WarIsACrime.org. He is a 2015 Nobel Peace Prize Nominee.

WILLIAM THOMAS resigned his U.S. Navy commission to protest the civilian slaughter in Vietnam and moved to Canada, where he became an award-winning reporter and photographer. In 1976, he spent eight years circling the planet in a hand-built trimaran and working as a freelance photojournalist, publishing his work in eight countries. He cofounded Canada's Green Islands Society and the Georgia Strait Alliance. During the Gulf War, he spent five months in the war zone as part of an ad hoc Gulf Environmental Emergency Response Team, an experience that lead to the book *Scorched Earth* and his award-winning film, *Eco War* [https://www.youtube.com/watch?v=yrR6cGU7DUA]. Thomas is the author of ten books, including *Bringing the War Home, All Fall Down,* and *Days of Deception.* http://willthomasonline.net/

KLAUS TOEPFER served as Germany's federal minister for the Environment, Nature Conservation and Nuclear Safety before being appointed under-secretary-general of the United Nations. He served as executive director of the United Nations Environment Programme from 1998 to 2006. He was the founding director of the Institute for Advanced Sustainability Studies and currently serves on the advisory board of the German Foundation for World Population. He also heads DEMOENERGY, a project dedicated to "the transformation of the energy system as the engine for democratic innovations."

DOUG WEIR is the project manager of the Toxic Remnants of War Project, a unique British-based research organization that monitors the impacts of war, military operations, and munitions on the environment and human health. TRWP's website serves as a global resource for policymakers, humanitarian organizations, and members of the public concerned with mitigating the effects of war on victims and communities. More information is available at www.toxicremnantsofwar.info.

JODY WILLIAMS has directed the International Campaign to Ban Landmines since its inception in 1991. Prior to joining the Landmines Campaign, she was first director of the Children's Project and the associate director of Medical Aid for El Salvador. In 1997, she was awarded the Nobel Peace Prize for her work with the ICBL.

S. BRIAN WILLSON is a Vietnam veteran and a trained lawyer whose wartime experiences transformed him into a nonviolent pacifist. On September 1, 1987, he was nearly killed during an antiwar protest when he sat down on the rails to block an ammunition train. The train severed both his legs. He has continued to demonstrate for peace, marching on two prosthetic legs and commuting by a special hand-powered tricycle. He is the author of *Blood on the Tracks: The Life and Times of S. Brian Willson*.

WIM ZWIJNENBURG is a Humanitarian Disarmament Project leader for PAX. His works focuses on conflict and environment in the Middle East and the use and proliferation of emerging military technology. KRISTINE TE PAS was an intern at PAX supporting the Conflict and Environment Program and is currently a consultant on corporate Immigration.

ACKNOWLEDGMENTS

Having spent much of my life as an environmental advocate and an antiwar activist, I appreciate the value of coalition building. That's why I cofounded Environmentalists Against War—to forge an alliance between pacifists and ecologists. So it was a deep personal pleasure to assemble this collection of essays on war and the environment. But this book wasn't my idea.

It was Douglas Rainsford Tompkins—indefatigable adventurer, mountain climber, river rafter, pilot, pioneering entrepreneur (the founder of North Face and Esprit Clothing)—who first saw the need for this book. As the visionary force behind the Foundation for Deep Ecology (FDE), Doug envisioned a large-format, full-color collection of essays and photos that would stir the mind and stun the eye. He had done this before, in a series of powerful advocacy books that include *Clearcut: The Tragedy of Industrial Forestry* and *Energy: Overdevelopment and the Delusion of Endless Growth*.

By December 2015, the book was well under way, thanks to the FDE project crew that included Doug, Tom Butler, George Wuerthner, and Daniel Dancer. Tragically, that was the month Doug succumbed to hypothermia following a kayaking accident on Chile's General Carrera Lake. With Doug's death, it became necessary to focus on the immediate needs of Doug's philanthropic work, and the book project had to be suspended.

Thanks to Helena Cobban and the team at Just World Books—Brian Baughan, Steve Fake, Marissa Wold Uhrina, and Kristin Goble—this legacy project has found a new home and will, hopefully, reach many readers. My deepest thanks to the many authors and publishers who provided original essays and updated previously published articles for inclusion in *The Reader*.

During a memorial service for Doug in San Francisco, Wes Jackson observed: "If your life's work can be accomplished in your lifetime, you're not thinking big enough." Doug dreamed big. Working alongside his wife, Kristine McDivitt Tompkins, and joined by various partners—including several presidents of Chile and Argentina—Doug helped create five new national parks (including Monte León National Park and Yendegaia National Park) and expand another, protecting more than three million acres in vast wilderness reserves. Since his death, the Tompkins Conservation team has collaborated with Argentine president Mauricio Macri to establish the new Iberá National Park, and is working to create at least five more national parks in the coming years. More info at www.tompkinsconservation.org.

Doug Tompkins had a simple explanation for his campaigns to protect the wild places of the Earth: "It's the rent I pay for living on the planet."

When it comes to putting an end to wars that now threaten to put an end to life on Earth, that's a fine mission statement for us all.

PERMISSIONS

Grateful acknowledgment is made to the following authors and publishers whose copyrighted works have been reprinted or adapted for use:

"Enheduanna's Lament." Translation by Daniela Gioseffi. From *Women on War, an International Anthology of Writings from Antiquity to the Present*, edited by Daniela Gioseffi (The Feminist Press of the City University of New York, 2003).

The Roots of War

"U.S. Exceptionalism: The Hubris That Fuels Wars" is adapted from *Drones and Targeted Killing: Legal, Moral, and Geopolitical Issues*, © 2014 by Marjorie Cohn, used with permission of the author. www.consortiomnews.com.

"Blowback: Climate Change and Resource Wars" is adapted from "Entering a Resource Shock World: How Resource Scarcity and Climate Change Could Produce a Global Explosion," © 2013 by Michael T. Klare, used with permission of the author and *TomDispatch*. www.tomdispatch.com.

"War Is Not Biological" is adapted from "Warfare Is Only an Invention—Not a Biological Necessity," © 1940 by Margaret Mead. Reproduced by permission of the American Anthropological Association. Not for sale or further reproduction.

"Patriarchy and War: Treating Nature Like Dirt," © 2017 by Vandana Shiva.

"Lessons from the Bonobos" is adapted from "Bonobo: Messenger of Peace, Victim of War" by Sally Jewell Coxe, © 2002 by the *Animal Welfare Institute Quarterly*. Used by permission of the AWI and the Bonobo Conservation Initiative. www.bonobo.org.

The Business of War

"War Is a Racket" is adapted from *War Is a Racket*, © 1935 by Major-General Smedley Darlington Butler. Published by Round Table Press, New York.

"War as an Economic Strategy" is adapted from *The Capitalism Papers: The Fatal Flaws of an Obsolete System*, © 2010 Jerry Mander, used by permission of the author and Counterpoint Press. www.counterpointpress.com.

"An Empire of Military Bases" is adapted from "Empire of Bases," © 2009 by Hugh Gusterson, used with permission of the author and the *Bulletin of Atomic Scientists*.

"The High Cost of a Warfare Economy" is adapted from "Don't Tread on Me," © 2014 by Dave Gilson, used with permission of the author and *Mother Jones* magazine. www.motherjones.com.

"Disaster Capitalism on the Battlefield" is adapted from "Disaster Capitalism on the Battlefield and in the Boardroom," © 2013 by William J. Astore, used with permission of the author and *The Washington Spectator*. washingtonspectator.org.

" 'False Flags': How Wars Are Packaged and Sold," © 2016 by James Corbett, courtesy of the Corbett Report. www.corbettreport.com.

"Banking on the Bomb: Investing in Nuclear Weapons" is adapted from *Don't Bank on the Bomb: A Global Report on the Financing of Nuclear Weapons Producers* by Susi Snyder and Wilbert van der Zeijden. © 2016 PAX and the International Campaign to Abolish Nuclear Weapons. www.dontbankonthebomb.com.

"Recruiting America's Child Soldiers" is adapted from "America's Child Soldiers," © 2013 by Ann Jones, used with permission of the author and *TomDispatch*. annjonesonline.com.

Nature in the Crosshairs

"Afghanistan: Bombing the Land of the Snow Leopard" is adapted from an article in *CounterPunch*, © 2010 by Joshua Frank, used with permission of the author and *CounterPunch*. www.counterpunch.org.

"Africa: Wars on Wildlife" is adapted from several articles by Dr. Jane Goodall, including "Devastating the Earth: The Unseen Victims of War," *Resurgence*, May–June 2003. Used with permission of *Resurgence* magazine and the Jane Goodall Institute. www.janegoodall.org.

"Guam: The Tip of America's Global Spear" is adapted from "US Military Bases on Guam in Global Perspective," © 2010 by Catherine Lutz, used by permission of the author and *The Asia-Pacific Journal*. http://apjjf.org.

"Kuwait: The War That Wounded the World" is adapted from "Eco-Activists on the Ground in Kuwait" by William Thomas (© 1991 *Earth Island Journal*) and *Scorched Earth: The Military's Assault on the Environment*, © 1994 by William Thomas, used with permission of the author and New Society Publishers. www.newsociety.com.

"El Salvador: Scorched Earth in Central America" is adapted from "El Salvador's Invisible War," © 1988 by Gar Smith. Used by permission of the author and *Earth Island Journal*.

"Serbia: The Impact of NATO's Bombs" is adapted from "What NATO's Bombs Did to the Environment" by Phillip Frazer, © 2000 News on Earth Ltd. Used with permission of the author, News on Earth, Ltd., and *Earth Island Journal*. www.earthisland.org.

"Sardinia: Bombs and Cancer in Paradise" is adapted from "Sardinia: Militarization, Contamination & Cancer in Paradise," © 2012 by Helen Jaccard, used by permission of the author and WarIsACrime.org.

"Vietnam: Delivering Death to the A Luoi Valley" is adapted from *Month of Pure Light: The Regreening of Vietnam*, © 1990 by Elizabeth Kemf, used with permission of the author and The Women's Press. www.womenspress.com.

Collateral Damage

"Baghdad: A Civilization Torn to Pieces" is adapted from "Iraq: A Civilization Torn to Pieces," © 2003 by Robert Fisk, used with permission of the author and the *London Independent*. © 2003 Independent Digital (UK) Ltd. www.independent.co.uk.

"Ukraine: Civil War and Combat Pollution" is adapted from a blog originally prepared for sustainablesecurity.org by Doug Weir of the Toxic Remnants of War Project in Manchester, U.K., with contributions by Nickolai Denisov and Otto Simonett of the Zoi Environment Network in Geneva and Dmytro Averin of the East-Ukrainian Environment Institute in Kyiv—Slovyansk. www.toxicremnantsofwar.info.

"Syria: Cities Reduced to Toxic Rubble," is adapted from an October 2015 report, *Amidst the Debris*, an eighty-five-page investigation by the Netherlands-based organization PAX. www.paxforpeace.nl.

"Wars and Refugees" is adapted from *Ecology of War and Peace: Counting Costs of Conflict* by Tom H. Hastings. University Press of America (2000). Used by permission.

"Civilian Victims of Killer Drones" is adapted from *Drone Warfare*, © 2013 by Medea Benjamin, used with permission of the author and Verso Books.

"The Navy's Sonic War on Whales" is adapted from "Lethal Sounds: Deadly Navy Sonar Harms Whales," © 2014 by Pierce Brosnan, used by permission of the author and the Natural Resources Defense Council. www.nrdc.org.

A Field Guide to Militarism

"The Militarization of Native Lands" is adapted from an essay that first appeared as part of *The Militarization of Indian Country* by Winona LaDuke and Sean Aaron Cruz, published under the "Makwa Enewed" imprint of Michigan State University Press, © 2013.

"Wars Are No Longer Fought on Battlefields" is adapted from *War Is a Lie*, © 2010 by David Swanson. Used with permission of the author and Just World Books. www.justworldbooks.com.

"War on Land: Toxic Burdens and Military Exercises" is adapted from *The Ecology of War: Environmental Impacts of Weaponry and Warfare*, © 1993 by Susan D. Lanier-Graham and used with permission of the author and Walker & Co.

"War on the Sea: Islands under Siege," © 2017 by Koohan Paik.

"Wars for Sand: The Mortar of Empires," © 2016 by Denis Delestrac.

"War at a Distance: Long-Range Missiles" is adapted from "PMTR: Kaua'i's Biggest Bang Is Out of Sight" © 2014. Used with permission of the author and *The Hawaii Independent*. http://hawaiiindependent.net.

"War in Space: Astro-Imperialism" is adapted from "Astro-Imperialism: War in Space," © 2002 by Karl Grossman, used with permission of the author and *Earth Island Journal*.

The Machinery of Mayhem

"Fueling the Engines of Empire" is adapted from *The Green Zone: The Environmental Costs of Militarism*, © 2009 by Barry Sanders, used with permission of the author and AK Press. www.akpress.org.

"Superpower, Superpolluter" is adapted from an article by the same title, © 2000 by Tyrone Savage, used with permission of the War Resisters League. www.warresisters.org.

"Jet Fright: The Impacts of Military Aircraft" is adapted from an article of the same title, © 1989 by Petra Loesch, used with permission of the author and *Earth Island Journal*. With research from the book *Tiefflieger* by Olaf Achilles and Jochen Lange.

"Explosive Arsenals: 'Bomblets' to 'Near-Nukes.'" Excerpted from *The New Nuclear Danger*, © 2002, 2004 by Helen Caldicott. Reprinted by permission of The New Press. www.thenewpress.com.

"Land Mines: The Smallest WMDs" is adapted from "Landmines and Measures to Eliminate Them," an essay by Jody Williams in the *International Review of the Red Cross* vol. 35, no. 307, published in August 1995. The article also includes portions of the author's 1997 Nobel Prize acceptance speech, © 1995 by Jody Williams. Used with permission of the author and the IRRC. www.icrc.org/en/international-review.

"Weather as a Weapon: 'Owning the Weather in 2025'" is excerpted from a June 17, 1996, U.S. Air Force Research Paper posted by the Federation of American Scientists.

"Nuclear Doomsday" is adapted from "The American Doomsday Machine," © 2009 by Daniel Ellsberg, used by permission of the author and TruthDig. www.truthdig.com.

The Aftermath

"Graveyards, Waste, and War Junk" is adapted from *The Ecology of War: Environmental Impacts of Weaponry and Warfare*, © 1993 by Susan D. Lanier-Graham and used with permission of the author and Walker & Co./ Bloomsbury Publishing.

"The Pentagon's Toxic Burn Pits" is adapted from "Ring of Fire: Why Our Military's Toxic Burn Pits Are Making Soldiers Sick," © 2013 by Katie Drummond (read the full story at www.theverge.com/2013/10/28/4771164/the-next-agent-orange-why-burn-pits-are-making-soldiers-sick). Used with permission of the author and Vox Media. www.voxmedia.com.

"America's 'Downwinders,' " © 2017 by James Lerager.

"Atomic Islands and 'Jellyfish Babies' " is adapted from "Nuclear Testing on Human Guinea Pigs," © 2013 by Darlene Keju-Johnson, used with permission of Giff Johnson and CreateSpace Independent Publishing Platform.

"Haunted by Memories of War" is adapted from *Blood on the Tracks: The Life and Times of S. Brian Willson* © 2011 by S. Brian Willson, used with permission of the author and PM Press. www.pmpress.org.

"The War Zone That Became a New Eden" is adapted from "Korea's Green Ribbon of Hope: History, Ecology, and Activism in the DMZ," © 2012 by Lisa Brady, used with permission of the author and *The Solutions Journal*. www.thesolutionsjournal.com.

Ecolibrium

"Take the Profit Out of War," © 1935 by Major-General Smedley Darlington Butler. Published by Round Table Press, New York.

"Ecology and War," by David Brower and Friends of the Earth. Reprinted with permission from *Not Man Apart*, October 1982.

"Why We Oppose War and Militarism." Press release from Environmentalists Against War, 2003.

"In Defense of the Environment," is adapted from "In Defense of the Environment, Putting Poverty to the Sword," © 2011 by Klaus Toepfer, used with permission of the author and the United Nations Environment Programme. www.unep.org.

"Protecting Nature from War" is adapted from "Protecting the Environment during Armed Conflict: An Inventory and Analysis of International Law" (November 2009), United Nations Environment Programme. www.un.org/zh/events/environmentconflictday/pdfs/int_law.pdf.